A Must-read for
Every Aspiring
Entrepreneur

12

Principles for Success

in Launching Your New Business

A STEP-BY-STEP GUIDE
To Help You Transform
Your Business and
Product Ideas
into Reality

How to Develop, Fund, and
Pitch Your New Business

Sam Hijazin

Published in the United States by Prodision, LLC

This book may be purchased in bulk for educational, business, or sales promotional use. For information, please email book@prodision.com

ISBN: 978-1-7373906-0-2 (Paperback)
ISBN: 978-1-7373906-2-6 (Hardback)
ISBN: 978-1-7373906-1-9 (eBook)

Library of Congress Control Number: 2021911641

Printed in the United States of America

First Edition – July 2021

⟨PRODISION

Allen, Texas

www.prodision.com

To my wife, Zoya, who makes my life
full of meaning, love, and joy.

Contents

Why Do Some Become Winners?

Part One: Pursuing the Great Idea

Part Two: Preparing for Success

Part Three: Transforming your Business Idea into Reality

Part Four: The Strategy for Making Money

Part Five: Time for Action

About the Author

Sam Hijazin is an accomplished executive and entrepreneur with more than 25 years of experience in developing and launching companies, products, and services globally. He has worked with Fortune 100 companies and launched innovative startups. His experience includes working with Verizon, AT&T, American Tower, Comsearch, Telcordia (Part of Ericsson), and Citigroup. He has also launched several startups and worked with accelerators, and venture capital-backed companies. He started his career as an engineer working in the Washington D.C. metroplex area designing, deploying, and launching the first commercial digital US wireless networks in the mid-1990s. He then grew his career to follow his passion for leading the launch of innovations and cutting-edge products and services. He led the development of new markets, and managed new product rollout and business growth in the mobile and wireless industry, in digital media and marketing, as well as in retail, financial, and ecommerce. He has worked across global markets including Europe, South America, North and Central America, Asia, the Middle East, and Africa.

Sam is also an in-demand speaker at several global conferences. His passion, work ethic, and skills have enabled him to engage with executives in the largest global companies in the world, yet his detailed knowledge and analytical abilities have allowed him to work one-on-one with all layers in these organizations, including engineers, scientists, software developers, marketing, sales, legal, risk, compliance, public relations, communication, and field employees. He has been able to offer leadership, guidance, and solutions to complex problems that face organizations' internal operations, as well as addressing customers' pain points and product growth opportunities. Sam has a Bachelor of Science in Electrical Engineering (B.S.EE) degree from the University of Alabama, and a Master's degree in Business Administration (MBA) with a focus on strategic management and innovation from Illinois Institute of Technology in Chicago.

Acknowledgement

To my parents, who seeded the principles of a well-lived life in me which became the foundations of my own. Their teachings are embedded in my life choices, including my education, work, and raising my family. I have watched you walking through life upholding your principles, and committing to applying them with responsibility regardless of how hard life got. Most importantly, I saw how rewarding your genuine and truthful relationships with others have been. Your kindness has paid off a lasting dividend in love and respect from all those you've come to befriend or work with. Thank you for inspiring me to become who I am, to continue seeking a meaningful calling in life, and to contribute wherever I am.

Author Preface

In 2001, after working for a few years as an engineer, I got the urge to learn more about the business side of work, beyond the science and technology. I wanted to gain new skills to complement my electrical engineering degree and prepare myself to take on new product innovation, not only from ideation to reality, but also to market. So, while living in Chicago, I pursued an MBA degree that was structured for those who already had a few years of technical experience and wanted to realize such a goal. I was partly driven by a desire to participate in the revolutionary new world of the internet, the exponential gain in computing processing power and speed, and the development of sophisticated communication infrastructure as led by the introduction of the modern personal wireless network. I wanted to develop the next mega unicorn company and, at the time, it felt as if there was a new one born every day, each one introduced and presented as the next Microsoft that would change the world as we know it. While many made the leap, such as Google, Netflix, Amazon, and Salesforce, and some others were acquired by bigger players, the majority of those new companies failed. Many had exceptional potential, yet they were never able to realize it.

During the MBA program, a statistic was presented to me early on by a professor. It left me in disbelief. Only 10% of all startups survive, and less than 1% actually grow to produce a meaningful revenue size. Coming from an engineering background, where every problem faced must be scientifically analyzed and pursued to reach a defined boundary and solution, such a statistic astonished me. What was more astonishing was that many of those businesses that failed were actually started by engineers. This stayed with me. Why did some startups win, while many others lost?

We now live at a time when there is a need for more successes than failures. The prospect of building a business that can last is so important to one's family, the city we live in, and the global economy overall. Starting a new business or new product, when successful, drives happiness, satisfaction, and a sense of accomplishment for yourself and every member of the team. It also offers an opportunity for independence and financial freedom. Successful businesses create jobs and drive growth and prosperity in the local and global economies. Many innovations that we enjoy today started with great ideas and passionate founders surrounded by the right advisors and skilled teams that helped propel their concept to reality. In the US, and throughout the world, I see passionate founders every year attempting to create the next great innovation that one day may become the next Apple, Tesla, or Facebook. Many are looking to advance innovations in the healthcare industry to defeat cancers or be more prepared to tackle the next pandemic. Some want to eradicate hunger and offer a cleaner and healthier environment or a less polluted planet for us and future generations. Those goals are important to me as well, and I'd like to see these pioneers succeed for the betterment of our societies.

As I progressed in my career, I was fortunate to be part of many technological revolutions and was trusted to lead the launch of innovative products within Fortune 500 corporates and found my own startups. During that time, I, too, became part of those statistics. Some of my work has succeeded and left a visible footprint, while other ventures have failed painfully. Through this process, I became one of the leading wireless engineers who collaboratively helped launch the personal communication system in the US, working side by side with US mobile carriers addressing the complexity of rolling out new wireless technologies and networks in many of the major markets in the US. I have also led critical digital transformations in the financial industry that created significant increases in value to those institutions.

Throughout this education, I learned that the process of building businesses relies heavily on applying principles. Consider how we apply careful protocols in designing critical structures, such as bridges that last more than 100 years, aircrafts that can fly at any elevation and in any weather conditions around the globe, and medicines targeted at the molecular level that can cure diseases. None of these are developed on instinct. By following clear principles that will make your business last, the chances of its success are substantially increased.

Today, many corporations, even with a wealth of cumulative knowledge and experience, still get it wrong more often than not. Although failures can be valuable learning opportunities and contribute to future wins, most new business founders cannot afford the time, nor the capital, to support avoidable mistakes. The need to get it right with the least number of tries, and fast, is critical to success and future growth.

My intention in this book is to offer you the principles to succeed in starting an innovative company or launching a new product; to hit the ground running and avoid the mistakes that you can ill afford to make. At the end, I provide you with a template that will guide you in constructing your business plan and strategy leveraging these principles. The template provides an easy-to-use, step-by-step structure that will put you in a much stronger position to meet your business goals.

I look forward to the day when that statistic is reversed, and the majority of startups are winners. I wish to see more businesses flourish and to help you realize your greatest possibilities. If you've started your company or developed a new product or service utilizing these principles, I'd love to hear from you. Please reach out and let me know how you did.

Good luck!

Sam Hijazin
sam@prodision.com

Why Do Some Become Winners?

Introduction

We live in a society that is obsessed with winning no matter what. We are told to never give up, to be persistent, to pick yourself up when you fall and dust yourself off, and that failure only makes you stronger. However, statistics shows that, on average, more than 50% of new small businesses fail in the first five years, and 80% to 90% of new startups fail within the first one to two years. One may conclude that we are a society that is more obsessed with failing than winning. We know this is not what we hope for when starting a new venture, yet it is often the outcome.

There is no shortage of inspirational material — articles, blogs, books, digital media, and movies -that embody these ideals. Indeed, they are great lessons that are absolutely inspiring and which we need to incorporate into our journey. We all, at one point, need to be reminded that failure is only a step in a process that can lead to success. However, the shelves are relatively thin when it comes to transforming these motivational quotes into actual, realized success. It is crucial to build knowledge and discipline in order to guide your passion. Getting inspired is fundamental to our success, but it is only effective when coupled with a practical roadmap that drives your ambition forward and initiates bold steps toward realizing your goals. This is what this book aims to do – to help you transform your dreams into reality by anchoring your aspirations to solid ground and proven techniques that will provide substance and a robust foundation.

If you're reading this book, you are likely someone who has a genuine interest in increasing your potential for success. You may also want to provide sound business advice to those who seek it from you, whether at work to your subordinates, management, and colleagues, at home to your family, or at a social event to your friends.

Why is this book needed now? At a time when innovation in corporations, small businesses, and startups is greatly needed to navigate the uncertainty imposed by the pandemic, and when there's an extremely competitive environment, as well as financial crises, following simple principles and processes can significantly improve the outcome of your efforts. While big businesses continue to drive innovation by allocating huge investment to research and development, and by introducing new products and services, there is an unprecedented number of new small businesses and high-tech startups that are sweeping the market, driven by developments in tools and resources available to us at a relatively cheap cost. The significant growth in digital transformation has made technology and information more accessible to new ventures, and concepts can be tested at a lower cost than ever before. The surge in startups is partly driven by a rise in those seeking independence and freedom from the traditional work lifestyle, especially in the younger generations, but also by many experienced employees, executives, and investors who want to become more autonomous. They see themselves as having the capability and knowledge to initiate their own ventures; ventures which may one day lead to something bigger and more impactful on society, the environment, and their own success.

Therefore, the need for this roadmap cannot be underestimated. The 10% to 20% success rate will not only leave many in society deprived of financial support but will also shut down global growth prospects and the potential for recovery which starts at the local microlevel; it also deprives us of innovations that can benefit society as a whole.

So, how will this book help you? By asking the important questions and avoiding false starts, a business can increase the chance of meeting its goals and securing its future. The principles I outline are simple, but we often either ignore them out of passion, wanting to launch the product too soon, or simply because we lack the discipline to exert the needed effort. Both can blind us to what can truly make our business grow and, more importantly, seeing what can actually make it fail. This is a painful outcome if it has great potential and a lot of investments shoring it up. The principles are intended to be implemented by business leaders, founders, managers, employees, and self-starters; those who seek to influence the success of the organization they work for, or own. From the chief executive officer to the product manager, the sales and marketing to the engineer, and from the first line of support to administrators. All those who support the business, in any form, can use such principles as a tool to guide them and their efforts in reaching a collective goal.

I want to make it clear that it's not a fear of failure that should be the motive here; rather, it is understanding the factors that can deprive a business from winning. Fear of failure can be a strong motivator. No one enjoys the loss of investment, income, and, more importantly, hope that comes with it. However, the emphasis here is that, if you ask the right questions combined with a sound business plan, a venture can be more prepared for the challenges and risks ahead, and able to address them in the early stages of the process.

Making mistakes is encouraged; it can be a great learning opportunity. The approach being discussed here is not intended to discourage the taking of risks; on the contrary, the intent is to help business leaders to venture out and possibly take more risks than usual because they are on a stronger footing. By simply asking and answering the critical questions, the chances of failure are reduced, and opportunities can be sought more aggressively. Whether you are leading a high-tech artificial intelligence (AI)-based venture, a hardware or software-based company, or launching a retail business, being prepared can focus your efforts to grow in the right direction. In the event that failure cannot be avoided, the learning experience should be maximized. With a firm foundation, you should be able to recover and become a stronger leader who can pivot, correct, and grow the current business or the next venture.

This guidance is based on three decades of practical experience working with Fortune 100 companies, introducing new products globally, setting up new services, and launching startups and new businesses from the ground up. It not only utilizes practical experience and skills built through my professional working years, but also relies on academic studies typically taught as part of a Master's in Business Administration (MBA). My goal here is to make this learning available to the largest possible base and help more businesses to succeed. The entrepreneur, the executive, the employee, and the student can benefit from following a disciplined path when developing a product or launching a business. For those who are already practicing these techniques, congratulations: you know how rewarding it is. For the rest of us, it is a great, practical guide and a path toward transforming our efforts into a sustainable reality.

What is the Secret Sauce of Success?

So, what makes some win and others lose? Often, when recognizing a successful venture, one may ask, "What is the 'secret sauce' that propelled this business to success?" Indeed, in every spectacular success story, there is a magical combination of factors that is imbedded in the DNA of the company. Whether it was intentionally developed, naturally existed, or was stumbled upon by accident, such a catalyst cannot be ignored.

Such secret sauce is often the cause of why two companies in the same industry can diverge so dramatically; one experiences great growth, while the other struggles to exist. It is an incorrect assumption that knowledge alone will produce the same outcome. The reason simply lays in the team who carry the company forward and their execution and implementation of the business principles and strategy. That is where the secret sauce is found. Put simply, individual's skills, experience, discipline, knowledge, culture, and passion are all factors that add to the uniqueness of the secret sauce and what propel a business to a distinctive destination.

Therefore, as you work through each of the principles, results may vary depending on how you apply them, the skills of the individuals involved, the time spent on each, and, most importantly, the decisions made in every relevant situation. These factors can make a significant impact on the outcome. The *execution* of each can be *your* secret sauce and your opportunity to differentiate yourself in the industry and among your customers.

Know and understand your secret sauce. Discover it as you go through the 12 Principles and decide where your strengths lie. Whether it is in you, in your team, or a combination of the two, let it shine in your execution and your decisions. If you are missing it, develop it, acquire it, or hire it.

The 12 Principles

Winging it, leaving it to luck, or following your emotions is not the sure path to business success. No business should begin without a plan constructed from the 12 Principles: it will bring order, logic, and rationale to your creative impulses; it will provide you with clear path forward based on your aims and purpose; and it will help you to articulate your objective worth and value to investors, customers, and the market as a whole.

The principles can be simple to follow once understood but require discipline to develop and apply. While they are all important for success, some of them may become more relevant than others, depending on your business offering and the stage that the company is at in its development. One thing is for sure: they should all be carefully considered and addressed when launching a business, a product, or a service, and should be revisited often as you optimize your plans, grow your market, or develop new offerings.

Why Follow the 12 Principles?

To understand the importance of following these business principles, one must understand the impact when such rules are not followed. For example, if a team involved in developing a multitenant residential building has followed the engineering code, but only partially, what impact will it have on the business itself and those involved in it? The foundations may not withstand the weight of the building, the structure of the walls may not handle the vertical stress or the environmental pressure, and the finished quality may not provide the insulation, the safety, or the overall usability and comfort expected by a resident. If you are invited to invest in such residential project, to buy a flat in it, or to move in to one of its floors, would you take that chance? Most likely, the answer from the vast majority of people would be a resounding no, possibly coupled with a horrified laugh. No one would want to be associated with such a project or use the product. It is simply destined to fall and fail.

We have seen many examples around the world of such devastating outcomes where businesses, without a sound design, fall and fail, with disturbing results; the impact on societies, corporations, individuals, and families is ruinous.

The Luck Factor

How does luck play a role in all of this? Often, luck occurs when an opportunity meets readiness. It comes unexpectedly and without significant effort. Otherwise, it is not really luck, it's hard work or smart decision-making that got you there. We've all seen businesses having good luck and success. Usually, this occurs because an environment has emerged that provides the right circumstances to support the business. This is truly what good luck is - when the opportunity presents itself and the business value, skills and structure happen to match that opportunity because of a set of unique conditions and timing. It may have occurred without significant research, analysis, or developing any readiness. While we've seen, or heard of, many examples, it is the exception. Let's look at an example. Jack buys a parcel of land offered by a friend where, later, a major commercial development is built adjacent to it and the land triples in value within a short time. This is certainly down to luck, without any forward planning. Another investor, Kate, may have bought an adjacent parcel of the same land, but she did it after significant research and market analysis, including carefully selecting the right location, determining the right price to pay, the size to buy, and how to pay for it. Both of those investors received the same amount of return from the rise in the value of the land; however, one got lucky, while the other reaped the results of well-calculated business decisions and sound capital allocation. The key difference, however, is that the one that followed sound business principles in making the investment is able to repeat this success in other investments they make over and over again, leveraging the knowledge and process that they followed the first time. While Jack may strike it lucky again, there are no assurances that future investments based on this random approach will achieve the same success. Some may have gotten by on luck so far, but it's unpredictable and unsustainable as a foundation for business success; the 12 Principles offer a strategy which is future-proof and which will enable you to repeat your success time and time again.

However, unexpected conditions may also occur, even if you follow disciplined business principles; luck can still have a hand in your success or failure. Sudden and unexpected changes in the market environment may result in a boost or a decline in sales, depending on the offerings you provide. For example, the company Peloton provides at-home fitness machines including a monthly subscription fee. They already had a great business model and targeted a specific segment of customers that prefer to work out from home and were able to afford the monthly subscription cost. When the COVID-19 pandemic hit, it impacted global markets; most gyms had to close for extended periods of time to reduce the possibility of virus transmission, and they stayed closed for several months. This event, while having a devastating impact on gyms and on those with membership, also created unique opportunities for a few businesses. Peloton now provided a highly desirable alternative for those who could no longer go to the gym and were looking for an equally intense workout at home. This led to dramatic customer growth, with exceptional quarterly results. In their *First Quarter of Fiscal Year 2021* report, which reflects the results as of the calendar month of September 2020, they ended with 137% year-over-year subscription growth to reach more than 1.33 million subscribers, and total revenue growth of 232% to $757.9 million. The opportunity presented itself and they were ready for it. Other businesses have also stepped up and provided products and services to utilize the opportunity and support the population during the pandemic. For example, there was rising demand for protective gear, including masks, disinfecting supplies, immune boosting supplements, and digital entertainment that can be streamed directly to your mobile devices and TVs at home, such as Netflix and Hulu. While some businesses experienced an increase in demand during the pandemic, many others experienced a steep decline due to reduced travel and restrictions on movement. For some, this meant a complete stop in commercial activities, and many have sadly gone out of business and filed for bankruptcy. Restaurants, movie theaters, retail malls, airlines, hotels, and all supply chain vendors that provide services to these businesses have suffered tremendously.

Therefore, luck, or misfortune, can certainly impact a business's ability to succeed and sustain its operations. It should be enjoyed when it comes and weathered when it gives you a beating, but it can never be depended upon as the basis for making long-term business decisions.

In producing a strategic plan, however, a business can prepare itself ahead of time for fortunate, or unfortunate, circumstances and develop a level of flexibility to respond to these unique conditions and swings in the market.

How Much Time Do I Need to Allocate to the Plan?

The principles are applicable for both small and large businesses. Asking yourself the *Leader's Success Questions* at the end of each principle can support your venture, regardless of its size. What may differ, however, is the time and effort required, which is directly correlated with the risk levels you're exposed to and the size of the opportunity. The plan may take just a few days to several weeks or more, depending on the complexity of the business and your existing knowledge. Some answers may be readily available to you or your team, while others may require further research, or external help to address. A small retail opportunity, for example, such as running a lemonade stand, where one may need to invest a small amount of funds to establish and operate for few days, will not require the same level of research and planning time as launching a multimillion-dollar retail business or a high-tech startup. A low-risk opportunity equals a quick plan, as the decisions are not as significant in scale and can be addressed easily. Spending a relatively small amount of money researching a low-risk venture may also provide a learning opportunity that can be used to optimize the business plan for subsequent iterations, something that will benefit the business as a whole and promote future growth. The lemonade stand business, for example, will benefit from relatively small efforts in researching the traffic level at various locations in the targeted area, the forecasted temperature of the days you intend to operate within, the availability of other alternatives (competition), the price offered by competitors, and the signage and messages to propagate and attract customers (marketing). This includes understanding the cost of the drink itself (financials), whether it is going to be freshly-squeezed lemon or a ready-mix powder (product) based on the customer segment, and the likely price to attract such a segment. Also, how to provide an image of cleanliness and follow health guidelines will need to be explored. Does the business require a license to operate? And, finally, if you fail, what is the total risk? A business that intends to serve parents and children while at a park on a hot day will certainly benefit from following such business principles to inform decision-making, even if it is a simple process that takes an hour or two. It would be ridiculous, however, to spend months conducting costly research before committing to a business that will only run for a few days with a relatively small projected revenue.

I have been part of planning efforts that some took several months to complete. Many of the telecom mobile companies have a typical six- to 12-month or more planning cycle for offering new mobile capabilities, plans, or features to market. Amazon has a similar planning cycle when offering major new services. For example, in late 2020, Amazon introduced Amazon Pharmacy to its customers. It may look like a simple offering to add to the rich Amazon platform, but it likely took significant time and effort to introduce, not only to plan and build the required software capabilities for customer engagement, but also to establish the right strategic business relationships with drug companies, distributors, and insurance companies, including government institutions for licensing and regulations compliance. While I don't have exact figures, I do know that major services such as this one can have planning, preparation, and development cycles lasting at least two years, which may have included the acquisition of strategic companies that can support the venture.

What Are the Principles?

Most of the 12 Principles are presented in the form of a series of questions, which allow the business leader to research and explore the fundamentals of their business in order to construct a robust business plan. The key here is forming a response to the questions, not with pride and passion, but with an unbiased and objective approach based on actual data and a valid understanding of the business aspects being discussed. Impartial research is needed. Our emotions can sometimes blind us with excitement and hope, provoking answers that satisfy our ambitions for success, but which may not necessarily be supported by facts or reality. Throwing yourself fully into this process is important; these principles will accelerate the launch of your business in the right direction and avoid false starts as much as possible. Your responses will help increase your awareness of your business's strengths, as well as risks that can cause friction or failure. While markets can sometimes move in a direction where even the greatest preparation may not be sufficient to avoid disappointment, such readiness allows for failures, if they happen, to be quick, less expensive, and, most importantly, ones that we can recover from. The path to success will include many small steps; some will work to support your journey and go as planned, while others will cause setbacks and failures. But even these can work toward your business learning and thus support your subsequent success. The purpose of using these principles is to maximize the wins and reduce the number and size of your failures.

Below, I outline the 12 Principles that will support you in constructing your full business plan:

1. Define Your Opportunity
What Problem Are You Trying to Solve?

2. Understand Your Customer
Who Are They and How Many Are There?

3. State Your Business Mission
What Are Your Values and Goals?

4. Choose Your Team
Who Will Translate Your Mission into Reality?

5. Specify Your Product
What Are the Key Functional Components, the Requirements, and the Long-term Roadmap?

6. Identify Your Competition
What Are Their Strengths and Weaknesses and How Do You Compare?

7. Determine Your Competitive Advantage
What is Your Differentiator and Unique Edge?

8. Know Your Financials
What Are Your Cost, Pricing, and Revenue Structures?

9. Build Your Go-To-Market Strategy

What Will Your Initial and Full Market Rollout Look Like?

10.Secure Your Funding

What Capital Do You Need and How Will You
Find Investors?

11. Prepare Your Pitch

How Can You Convince Investors, Partners, and Customers
to Get on Board?

12. Evaluate Your Risk

What is the Likelihood of Failure and How Will You Prepare
for and Mitigate Your Business Risk?

Throughout the rest of this book, I will review each of these principles individually and in more depth. Each one ends with a set of targeted questions that will help drive your thought process, and I will provide expert guidance that helps you to research, address, and answer these as relevant to your business.

At the end of the book, I have provided a template to structure and direct your own business planning; this will help you to draft your responses to use in your next journey, product, or venture. You can also access a digital version of this plan on my website at www.sams12.com, which will provide you with an editable template for ease of use.

Part One

Pursuing the Great Idea

Principle 1: Define Your Opportunity
What Problem Are You Trying To Solve?

Principle 2: Understand Your Customer
Who Are They and How Many Are There?

Principle 1

Define Your Opportunity:
What Problem Are You Trying to Solve?

The maelstrom of ideas that bounce around our minds everyday are the true beginnings of innovation. We observe, we read, we communicate, and we engage with our surroundings; then a sudden feeling occurs, urging us to express a thought. These ideas are often formed out of a current need; a solution that addresses a "pain point" or an opportunity to make our lives better, safer, more entertaining, or comfortable. They may occur organically, in real time, during a relevant task; while, at other times, they just hit you unexpectedly while reading, watching a movie, driving, walking, falling asleep, and, yes, during a shower, as many of you would testify. They often sound really good in the moment, then the interest to pursue them slowly dissipates during the hours and days ahead. Many of those ideas do survive this period, however, and indeed many become a reality, producing successful products that we all use in our daily lives. While not every good idea survives, often due to lack of interest, poor execution, or other environmental factors, other "not so good" ideas *do* make it through and these often become a source of pain for the creator, the investors who supported it, and the customer who fell for it. Eventually, after a lot of money and time has been wasted, they fail.

Throughout my career, having worked with many innovative organizations, I directly experienced the huge number of scientific creations that sat on the shelves and never saw the light of day. Many of those prototypes, while great science, failed to demonstrate practical benefits to customers. At this stage, having been developed from idea to demonstrable concept, there had already been a significant cost outlay. Others were actually found to be excellent products but were just awaiting a business decision to move them to market. They perhaps needed funding or approval of the productization process. However, they too failed to make it because they stayed too long in the early stages awaiting action. Market demand had shifted in this time, which made them obsolete.

The common saying, "Ideas are a dime a dozen," is often used to emphasize the fact that hard work is what makes an idea a success, and not necessarily the idea itself. While I won't dispute the fact that perspiration and execution is what transforms a concept into a successful reality, the concept still has to be robust and address a current or future demand from customers. It must be perceived as important and valuable to them, and it must deliver what the target customers need or want. Yes, ideas are a dime a dozen, but not all ideas are equal in value and worth pursuing. Ideas that do not serve an important purpose to a sizable segment of potential customers are likely to fail, no matter how much effort is made. Be sure that your idea is relevant and useful to your customers and the hard work will eventually pay off. To validate your idea in this way, you must ask yourself key questions that can help you determine whether to proceed as is, to optimize the idea in order to increase its relevance to your customers, or not pursue it at all and refocus your efforts and capital in another direction.

Today, with the availability of research tools, the internet, and communication platforms, developing a good idea that will resolve complex problems has become more attainable. These tools have transformed many of the services that we use day to day, delivering a much easier and smoother experience. Meal delivery services, for example, are building on increased demand for more convenience. DoorDash and Grubhub, among many others, are meal delivery companies that have developed a market value worth billions of dollars, leveraging the demand by consumers for ordering any type of meal to their location using a single provider. Opportunities to fulfill a specific demand like this have become much faster and cheaper to establish and test. Banking has become mostly digital with the ability to deposit, transfer, and pay bills from your own home or office desk, where before it meant a trip that took time out of your day or lunch break. Business meetings, job interviews, and lessons are happening over web applications such as Zoom, without participants having to set foot on a plane or travel hundred and thousands of miles to meet in person. Transfer of knowledge has gotten much faster as well. No matter where you are today, you will probably have equal access to information from across the globe via the internet. Innovative advancements in artificial intelligence, detection and surrounding awareness technologies such as LIDAR, data processing, and digital marketing are allowing for growth in areas beyond convenience and with highly-impactful economic and environmental practices. The revolution in renewable energy, the availability of more efficient and higher capacity electric battery storage used in cars and other machines, the greater dependency on the need to stay connected anywhere and anytime leveraging the latest wireless technologies, and the increase in remote collaboration applications whether for work, education, or healthcare services are all producing waves of great new innovations and opportunities.

The success of a business is highly dependent on the relevance of your product or service to your customers. Identifying what pain point you're solving, and how important those problems are to your customer, brings needed validation to your business and your product idea.

Ask yourself the following questions to validate the opportunity you are pursuing:

1. What problem or need are you addressing?

In this first step, you must establish the opportunity being pursued. Typically, it is referred to as the "problems" you are trying to solve, or the "pain points" you intend to ease or eliminate for the customer. Having a clear definition of these will support subsequent steps in defining the solution and business strategy.

Let's look at Uber as an example. They are attempting to solve their target customers' pain point of not having easy access to a taxicab, not knowing when it will arrive, not knowing the cost of the trip ahead of time, and not knowing the condition of the vehicle, among others. They developed an operational business model and mobile app that specifically addressed and delivered these needs in a way that the average, traditional cab service could not and have thus become wildly successful.

Clearly defining the opportunity, the problem, or the need helps you to focus your solution and strategy on fulfilling it before launch. It's also critical to develop supporting evidence of this need and its urgency.

2. How urgent and critical is this problem to your customers?

Understanding how important the problem is will provide an indication of how likely it is that the customer will prioritize and engage with your business. **There are many problems that we face daily, but not all are worth addressing** as a matter of urgency or important enough to solve. The solution should have a measurable impact on our lives i.e., in terms of time, health, happiness, productivity, relationships, and/or environment. In terms of a business-to-business solution, it should have an impact on customer growth or improve internal operations for more efficient and productive use of resources and capital. Ultimately, you need to be able to prove this problem to be important to the segment (population and customer type) you are targeting.

When easing the pain point is vital to the user, then it is more likely that your offering will demand the desired attention as they are more willing to part with their time and money to obtain it. Solving a problem that is not urgent or does not provide a measurable improvement for customers will likely face a disappointing response from consumers already overwhelmed with choices and short on time. Choose the problem you are solving carefully as, no matter how great your business plan is, customers' perception of its value will always dictate their engagement and interest in your offering. Whether it is a "need" or a "want," both can provide a strong foundation for a business proposition, as long as they feel it is important to them.

3. What are customers doing now to address the problem you have identified? Why do you think a new solution is necessary?

It is important to understand how customers are meeting their needs today without your offering. They are possibly being satisfied currently by one of your competitors. This is a challenge, but you need to recognize and analyze it in order to find a way forward. **Your solution must provide a substantial improvement over what already exists**. It could be in terms of saving them time or money; providing better quality, support, and ease of use; or improving its safety or environmental credentials. It may even be that you are improving on your own previous offering. Typewriters were iconic devices that became a necessity for almost every business, educational institution, government office, and for some individuals like writers. After more than 200 years of evolution, in the 20th century it became an irreplaceable item. Then, in 1983, a software application was launched by Microsoft, called WordPerfect and, later, simply Word. Despite their emotional attachment to the historically successful typewriter and the training needed to use the new computer program, slowly the masses started to adopt this new way of typing. There was a shift from using the heavy, single purpose machine to a much lighter, simpler, and more convenient form. Microsoft Word offered a faster tool to transform your thoughts and express your writing with much less waste in paper and ink. You could review, delete, and rewrite in any style with much less effort, check your spelling, and "save" your work. Now, typewriters are mostly bought as ornaments. This is all to show that, no matter how much loyalty one may have to a specific device, tool, or business, innovations that resolve the pain points in our daily lives can make anything replaceable. When the value far exceeds the cost of change, eventually everyone who can afford it will move to adopt. Either way, choosing your new product or service must provide your intended users with a measurable advantage. We will address business differentiation in more detail in Principle 7. However, at this stage of the plan, you must recognize and acknowledge the significant and compelling benefits that you will be offering beyond what already exists.

4. What customer behaviors are you seeking to change or introduce? How are you changing these behaviors with your offering and how does that compare to the behavior needed to use existing solutions?

When you introduce your offering, what behavior changes are you expecting the customer will make? Does your product or service increase or decrease the number of steps needed to fulfil their need? Does the behavior change you wish to see require little adaptation by the customer, or does it require more significant effort to adapt to?

For example, with the Uber application, the target customers didn't need to adapt their behavior very much to use it; it was a simple switch. The user only needs a smartphone and a credit card for the service to work. The service provider, in this case the Uber driver, only needs to have a car in good condition that meets the Uber standards, a license to drive with a good record, and a smartphone. For the customer and the service provider, these requirements produce the minimum level of friction in order for the service to receive great engagement rates by users. Of course, if the application was more complex, user unfriendly, or required a high level of specific skills, then fewer customers would be willing to use it. The key to this type of business is to be ubiquitously available and useable for the majority of people in order to succeed. Other business types may only need a small and specific customer segment to adapt their behavior in order to use their product. Doctors, for example, are capable of learning new and complex procedures and so will be open to more adaptation of their behavior in order to use a new surgical tool. However, the required level of adaptation will still need to be justified by significantly improving the performance of current tools. Knowing the type of business which you are in, with respect to this adaptation question, is important.

In general, though, users will expect only a small level of behavior adaptation to perform the same task, even when greater value is expected. You need to consider the leap that the customer will have to take in order to use your product, and how willing they might be to put in the time and effort required.

Trying to balance innovation with an accessible and simple user experience is hard to achieve, but those that succeed will likely reap the rewards.

LEADER'S SUCCESS QUESTIONS

Principle 1: Define Your Opportunity
What Problem Are You Trying to Solve?

Ask yourself the following questions to validate
the opportunity you are pursuing:

1. What problem or need you are addressing?

2. How urgent and critical is this problem to your customers?
   Do you think they will perceive your business, product, or
   service as important to them and worth prioritizing?

3. What are customers doing now to address the problem
   you have identified? Why do you think a new solution
   is necessary?

4. What customer behaviors (operations, processes, or
   way of living) are you seeking to change or introduce?
   How are you changing these behaviors with your offering
   and how does that compare to the behavior needed
   to use existing solutions?

Principle 2

Understand Your Customer:
Who Are They and How Many Are There?

U nderstanding who your customer is and the size of
the market can help you design your product or service more accurately to meet their needs. Whether
it is an internal customer within the organization,
a consumer, or another business, a good characterization of the
target customer and their priorities is vital for strategy and business success. This is also helpful for estimating the true extent
of the opportunity. Jeff Bezos is the founder of one of the most
recognizable global companies, Amazon. When asked about the
single most important element that contributed to their success,
he always answers, "Being customer obsessed." This could relate
to many different things, but it certainly reflects a prioritization
of the customer as a driving force for the business. This is a very
simple goal, yet it can be extremely complex to define, implement,
and achieve. It requires that you be continuously attuned to your
customers' engagement and feedback and that you understand
when they express dissatisfaction and when they are pleased. It's
a continuous attempt to understand what is important for your
customer in order to retain their business, leveraging the data that

may get collected from their engagement. Amazon has learned over time what makes their customers tick and have optimized their service accordingly to ensure satisfaction. This has translated to fast shipping, free delivery, having a wide selection, using community ratings to recognize quality, ease of search, and competitive prices. Understanding who the customer is can help you to develop products and services with a customer-centric strategy, and to continuously adapt over time.

Ask yourself the following questions to explore these issues in the customer assessment section of your business plan:

1. Who are you trying to help? Who may want to use the business's offering?

Defining your customer in this way can help you to better attract them to your business. For example, is the offering geared toward businesses, the government, or consumers? This initial high-level assessment will help you drive your product with more focus to the likely segment or segments. If the offering is for consumers, is it for a certain age group, a particular persona, or those at a specific education level? Is it geographically sensitive and limited to specific areas, cities, countries, and is it used by specific audiences (e.g., the technology savvy) or a wider audience? It could target a segment that has a specific need, such as those requiring a certain diet (e.g., the gluten intolerant), those with a certain lifestyle (e.g., that travel regularly for business), or with certain healthcare needs (e.g., those with a chronic condition).

You may also be targeting a segment based on their purchasing power. Walmart's business model, for example, is built on providing low-cost products but typically with limited options. It is designed to appeal to a very large segment of the population who are price sensitive and prefer to find most of their shopping in one place. On the other hand, there are retailers whose business is structured toward those who are happy to pay for choice, customized options, and access to qualities that meet their own tastes. For example, a small boutique tailor who designs bespoke suits and dresses with a choice of fabric and color. Each approach can be extremely profitable and successful as long as you understand your target customers and structure your strategy accordingly. One of the most dangerous traps to fall into as a business is losing focus on your customer, which can make your product very confusing to your potential audience.

2. What is the initial customer segment that you want to target and acquire?

When starting a business, you may want to sell to a wide range of customer segments; the bigger the market the better, right? While this is true for an eventual target, it is usually more sensible to define an initial set of customers that will be friendly first adopters and focus on attracting them to your product over others. This will help your business to focus its energy in acquiring an initial customer base, which is useful in learning more about your offering, how well it is perceived in the market, and any adjustments that may need to be made before going to mass market or expanding to other segments. For example, if you provide a business service, you may select a handful of companies to participate in a free trial to work with you. They would then provide their personal feedback in exchange for a measurable discount if they decide to buy afterwards. You may have won an early adopter, a reference to spread your message, and their insight could offer great value in better optimizing your product for the target customer. They may agree to join you at industry conferences, webinars, and in meetings with other (non-competing) customers. Similarly, for consumers products, in addition to friends and family, you may start with a limited offering within your own online or physical store, or within specific retailers in a particular neighborhood, town, or city, to gain initial insight. You can offer benefits to the retailers, such as discounts or priority service, in order to obtain the feedback. The data collected will be invaluable in analyzing how easily customers can discover, buy, and access your products or services; where customers get stuck in making the decision or in actually obtaining and buying the product; and where they drop out during the buying process. Other measurements to track can include repeat purchases, revisits, customer praise or complaints, customer service calls, and online ratings and reviews. The purpose of the limited initial offering is to listen to the customer and learn from their direct, or indirect, feedback. In this way, you have the opportunity to optimize your offering before going to the masses.

3. How large is the market? How big is the opportunity within the market?

Now that you have defined your customer segment or segments and understand their needs (or wants), it is helpful to measure the overall market size for your product. This knowledge will help you quantify the full potential of the business and assess your financial assumptions and planning, as we will discuss in more detail in Principle 8 (Know Your Financials). It will also enable you to validate the investments that you intend to make in the business. For example, a retail business that is targeting an age group within a specific location can research this segment and conclude that there is a defined and measurable number of potential users who live in the target area and would have interest in buying the product. If the business is not as limited by geography, for example if it is an online retail business in America that can deliver, then you can measure the number of potential users who fit the described target segment within the larger boundary of a region or a country. Some offerings have no dependency whatsoever on political boundaries or geographies, especially if they're based on information technology that can be delivered digitally online to customers globally. In this case, the potential market size can be almost unlimited. We are not addressing governmental regulations and restrictions at this point, which can be limiting, even for digital products. The key here is to measure the overall likely market size that your product can participate in, and potentially capture a share of, with reasonable business efforts.

For example, a pediatric dentist serving families with children can determine the full market size by researching, in the city records, the number of families who live within five to 15 miles of the clinic's location and estimating the number of children they might have under the age of 18. A business that develops dentistry equipment and tools, in comparison, may be measuring the market size based on the total number of dentistry clinics within a given country or even a global region. Google, which is mostly in the digital information business, may measure its market size based on the total number of global users with a digital device such as a smartphone, a tablet, or a personal computer (PC), and who can have access to the internet through wired, or wireless, communication. While there are regulations that may restrict how they sell or reach certain countries, their product is accessible for the majority of the world.

It is important to mention here that estimating the "market size" for a business is not to be confused with estimating the likely "market share" that your business may acquire. While there are a few companies that have dominating positions of more than 30% market share, such as Amazon, Walmart, Apple, Microsoft, Flipkart, Tencent, Google, and Tesla, they are the exception and it is rare for a single company to own such a significant portion of the market. The market share is the likely share of customers that you might win from the total estimated available customers (market size). This is important to recognize when projecting your finances for the first one to five years. In the dentistry clinic example, they will be vying for customers against other dentists in the area who are sharing the same market. Obviously, the market share of a given business will depend on many factors that will be addressed throughout the book. These might be the quality of the offering, pricing, the location relative to the customer, ease of use, and competition. We will address the market share calculations in Principle 8 (Know Your Financials).

4. Is the target industry growing, stable, or declining? Are customers growing in this segment or declining?

You may have defined the market size, but it is important to realize that the world is in continuous motion and not fixed. The needs of consumers, businesses, and governments change over time. The rate and direction of this change must be recognized. Some needs change faster than others. Some products grow in demand, while others decline. For example, in the mid-90s, it became apparent that what is called the landline, or simply the home phone, which had existed for decades as the primary line of communication, had started to diminish in usage and value due to the rise of wireless communication and the demand for mobility, connecting people anywhere and anytime. Companies who recognized these changes early on have evolved to weather the storm; they continue to exist because they remained relevant, investing in wireless infrastructure, wireless network services, and mobile retail stores, such as AT&T, Vodafone, Verizon, and Reliance. Others are long gone as they ignored the major shift in customer behavior and stagnated. WorldCom, which was known as MCI, a dominating telecom long-distance calling provider in the US, and others such as Global Crossing, XO Communications, and Nortel, the largest Canadian Telecom equipment manufacturer, tried but failed to innovate quickly enough to stay competitive and relevant.

This evolution in customer needs generated yet another leap in progress, which changed customer behavior yet again. There was now a need for fast and high-capacity communication to support the large usage of digital wireless communication, which encouraged the creation of fiber optics and high-speed internet. This growth in data usage has also led to the need to store, manage, and access the data more efficiently, which led to the development of data hosting centers, and later evolved into what we know today as "the cloud." Cloud hosting and computing allows companies to install their applications on a common computing infrastructure. Businesses invested in building this infrastructure and now act as hosts, leasing it to others. The demand for this service has grown significantly because of the value it offers to companies of all sizes. Businesses no longer need to buy their own expensive computers and servers; rather, they can access the readily-available hosted infrastructure for a reasonable monthly subscription fee, and can scale at a much faster rate than if they were managing their own computing infrastructure internally. Amazon, through their Amazon Web Services (AWS), Microsoft (Azure), Rackspace, and others recognized this demand for shared computing facilities and services and large amounts of data early on. They invested heavily in providing cloud-hosting services with features that the majority of businesses care about, including significant security measures, scalable data capacity, processing power, and accessibility from anywhere, anytime, 24 hours a day. Understanding this trend and growth direction has helped those cloud-service providers to make the leap and invest and prepare for such a significant demand.

Understand your target customer, how big the opportunity is, and if the segment is in growth or decline. Companies can still make money in shrinking segments, but your business must acknowledge the market direction and establish a strategy and plan to address the opportunity accordingly. In the long run, however, for those companies that want to stay relevant, serving a growing industry has greater potential for success than serving a dying one.

LEADER'S SUCCESS QUESTIONS

Principle 2: Understand Your Customer
Who Are They and How Many Are There?

Ask yourself the following questions to help you explore these issues in the customer assessment section of your business plan:

1. Who are you trying to help? Who may want to use the business's offering?

    As applicable, specify the target segments such as age group, education, locations, interests, income level, and career type. If it is a business-to-business offering, then describe the industry, the functions, user skills, and role in the company. You should take the same approach if it is targeting the public sector or a government sector.

2. What is the initial customer segment that you want to target and acquire?

3. How large is the market? How big is the opportunity within the market? This describes the total market size and the potential market share which you can obtain.

4. Is the target industry growing, stable, or declining? Are customers growing in this segment or declining?

Part Two

Preparing for Success

Principle 3: State Your Business Mission
What Are Your Values and Goals?

Principle 4: Choose Your Team
Who Will Translate Your Mission into Reality?

Principle 3

State Your Business Mission:
What Are Your Values and Goals?

As you develop a comprehensive understanding of the problem you are solving or the opportunity you are seeking to capture, and you are able to define the customer, the market size, and how important or urgent your product is to them, you can now begin to describe the mission of the company. The mission must express the company's passion toward serving the customer. It should be expressed across the company to inspire your team and convey your commitment to your customers. Once defined, you should then establish a set of strategic goals or objectives that will guide the company to achieve its mission. The mission statement drives not only the development of your products, but also your marketing plans, advertising strategy, and your specific messaging to target customers.

For example, Microsoft had a simple mission statement when they started, which was:

"A computer on every desk and in every home."

They wanted to make the computer accessible to, and used by, everyone and not just limited to corporations, professionals, or computer scientists. By introducing the Microsoft operating system — Windows 1.0 — they made it usable for almost anyone. We must acknowledge, however, that the company took several years until it reached its first goal and went through multiple business cycles and challenges before it evolved to the point at which it could fulfil this objective.

Another good example of a clear business mission is Nike, which wants to:

"Bring inspiration and innovation to every athlete* in the world." (Then, they nicely add: "*if you have a body, you are an athlete.")

The statement is inspiring, but has clear objectives which become part of the company's goals and culture. Their mission is to inspire and make performing sports activities and physical exercise a possibility for all. It should not just be limited to professional or elite athletes, but open to any person regardless of their physical capabilities, size, fitness, or age. Such a mission drives the company to develop sports products that are suitable for the various segments of the population: the young, the old, athletes with extensive workout goals, and the "athletes" who desire a less intense program and prefer a walk in the park, for example.

Other sports companies' missions may be more focused on serving a specific sport or athlete segment. None of these strategies are incorrect as long as they convey, with clarity, the company's focus and define who you want to serve, and what value you'd like to offer the customer.

Speedo, for example, a company that is focused on swimwear, have a mission to:

"Inspire people to swim — and to bring them the swimsuits and gear that make every moment in the water better."

Having a clear mission statement supported by a defined set

of objectives drives focus toward the design of your product and the business as a whole. It will help you set clear milestones which, once achieved, will lead you to satisfy and deliver the company's goals and values.

It's important to share the mission statement and objectives with all those involved so they understand how to deliver your vision. The purpose of the company should be clear in the mission, which can motivate and inspire your team to achieve. Keep it visible to you and your staff at all times so it can guide your strategy, processes, and actions on a daily basis; achieving it will require your full attention. Some companies frame their mission and hang it on the walls of the organization; others make it appear on every computer screen as a reminder to all of what the company's values are and its commitment to customers and society.

Keep your Mission Attainable

Aim big and aim high, but plan to start small. Your mission should absolutely be as ambitious as you think it needs to be. But be mindful of your available resources, capital, and the time that you can allocate to achieve it. Whether it is to establish the best shopping center in your own town, or to develop a chain of shopping centers so that everyone in every continent can enjoy them, it is you and your passion that should decide what it will be. Great missions, no matter how big or challenging they may appear, can be achieved by breaking them down into individual parts, goals, and objectives. Each element can then be executed through strategic planning and via a roadmap. This is how most successful businesses reach their goals. Facebook, for example, started with a limited mission to serve students in one college, Harvard, and then expanded to other colleges before it became a worldwide phenomenon. Coca-Cola, and PepsiCo, both established their early missions based on wanting to make their drinks accessible to everyone in every corner of the world — significant missions for two local drinks companies. While their mission evolved over time to include a larger range of products, they continually focused on this goal and, indeed, succeeded in reaching just about every corner of the world, despite navigating challenging political landscapes and wars in the 19th and 20th centuries. Now they truly dominate their line of products, as anchored by the Coke and Pepsi drinks.

Some missions, however, require more capital and effort to begin. You need to be aware of the magnitude and scope of your goals. For example, establishing a new company that launches modernized space shuttles into space, that are more efficient than anything that's existed before, would require a substantial amount of time and money that only a handful of institutions in the world can afford. SpaceX, Blue Origin, and the few others which are attempting to solve this problem had the expectation that this mission required multiple years of investment and substantial resources to realize. Few have the network and resources of Elon Musk, Tesla and PayPal's co-founder, and Jeff Bezos, Amazon's founder, who are behind these ventures. In fact, they did not start their first ventures with such expensive and highly-complex missions. They started with innovative and ambitious missions, but they only required a modest level of capital which, once coupled with their passion and good business practices, meant they were able to launch their early ventures successfully.

There should not be any limitation to one's imagination or

creativity, but I do want you to understand the hard truths when deciding on your mission. You should assess the expected timeline, resources, and capital that is likely required to achieve it, as well as the complexity of the task.

LEADER'S SUCCESS QUESTIONS

Principle 3: State Your Business Mission
What Are Your Values and Goals?

Ask yourself the following questions to help craft your mission statement:

1. What is the core passion that you want to commit to addressing for your customer?

2. Who is the target customer or audience that will benefit from your efforts?

3. What is the solution that you will offer to fulfill your commitment to customers?
   Examine your mission statement and be sure you can set goals and objectives that you can feasibly achieve. Once you've established your mission statement, then define your objectives.

4. What are your business objectives and goals? Define the goals that, if satisfied, will support you in meeting your mission statement and fulfilling your commitment to customers.

Principle 4

Choose Your Team:

Who Will Translate Your Mission into Reality?

The leader, the team, and your network of talented and knowledgeable people translate your business mission into a successful reality. It's critical that every team member has a skill set that is relevant and important to their role and to the success of the business. Your network is an invaluable resource; surround yourself with trusted advisors and people that can add value to your venture, whether through product knowledge, market knowledge, strategy, or connection to target customers.

The Leader

When launching a business, or leading the development of a new and innovative product, you are the leader, and you will be for a while. The leader's ability to navigate challenges and respond with solutions can be a decisive factor in a company's success. Your timely strategic decisions, and your ability to communicate, both verbal and written, when working with other team members, partners, vendors, customers, competitors, and investors, can be the key to carrying the business forward in the desired direction. **Opportunities can be lost or gained depending on the soundness of your decisions made at every juncture and every critical milestone**. This is the moment that you can shine and set the company's direction, so make every decision count. Be mindful of every interaction; one well-chosen discussion with an anchor customer, or a shrewd decision on a product direction, could point the company toward a great future. Success occurs as a result of many factors, but if you are careful and make decisions as if your business life depends on it, then the accumulation of those actions will work toward your overall success. The wrong decision, the wrong comment, or the wrong email can all put the company on an irreversible course.

The leader should enable the team to express themselves and their thoughts freely toward meeting the objectives of the company. A committed and self-motivated team can achieve incredible results. The leader should provide guidance, direction, and correct their team when needed; they should also empower their team to meet and exceed their goals when ready. Teams are motivated by many factors. Money is certainly a motivating factor as an expression of appreciation, as it enables the employee to meet their life commitments. However, it has been demonstrated that the value of money as a motivator has limits, and more money does not necessarily equate to more productivity, creativity, or a greater work ethic. In fact, numerous studies have shown (and I can personally testify to this from my own experience) that employees are more motivated by factors that provide satisfaction through recognition, achievement, and the feeling of inclusion.

Employees are often motivated by the following:

a. Being recognized for their special achievements and
   hard work.

b. The opportunity to develop their skills and learn.

c. Feeling they are making a difference with a purpose.

d. Being able to participate in critical missions or assignments.

e. A clear career path with attainable growth.

f. Providing input toward strategic initiatives.

g. Being empowered to make decisions.

h. Flexibility in working conditions.

i. A culture that promotes trust, collaboration, and fun.

The leader can have a huge impact on these factors as they set the direction of the company, its culture, and its future. Empower your team and recognize their good work to gain their trust, commitment, and self-motivation. Solicit the advice of all that can help, then make your decision based on your own judgment. Be mindful, however, that the decisions you make carry the power to bring you closer to achieving your company's goals.

Advisors, Board Directors, and Friends

When establishing a business, having a number of good advisors and friends surrounding you can make a great difference to your venture. Advisors can help in providing input to your strategy and business plan, and in validating ideas around the product. They can also help you with introductions to their network and connect you with key early customers that are valuable and significant to launching your company or product. Their help can also extend to supporting your efforts in raising capital if you need external investors for funding.

Many advisors are willing to offer their help, particularly advice, for free. They may be close friends, or someone from the community that sees value in what you're doing, particularly if it helps a cause they are passionate about. It is a common practice, however, to offer advisors who intend to support you on a continuous basis, some equity or a discount on your company's shares as a gesture of appreciation. Often these advisors may also decide to invest in your company and become an early shareholder. There are regulations and rules that can dictate such transactions, which I address in Principle 10 (Secure Your Funding).

You may choose your advisors from your own network of friends, previous co-workers, or leaders you've worked with before. You may also get introduced to them through mutual friends, or investors who share a common purpose and goals in supporting you.

Great advisors can offer key strategic input that may help you optimize your business plans, your company's pitch, and product direction. They can help you in establishing needed relations with new customers and investors, so make sure you leverage their reach and connections. But be aware that not every advisor can benefit you equally. It is you who has to decide how to use their advice and contacts. Assess who truly helps you, keep them close, and offer to compensate them for their time so they are incentivized to participate in the success of the company.

The Team: Management, Business, Development, and Operations

The team, including executives, managers, engineers, and first line employees, are all behind developing the core competency of the business. They are the main driver of success or failure. Some businesses require a small team to launch, others require more. Their know-how, skills, education, experience, and cultural fit can be the determining factors for business growth. The skill required to develop and lead a team to deliver the offering cannot be underestimated and requires experience and capital to fund it. When launching a business, the need for competent resources is critical. Such resources may not always be widely available, and can be expensive to acquire, but are worth the effort. For technical and engineering resources, for example, having a dedicated team that is available to offer the needed skills at any time, and work on optimizing the product, provides a great advantage. Therefore, be selective and develop a skills metric that you can measure against when acquiring resources and building the team. Focus on the roles and what associated skills are needed to be most successful in ensuring that the product and business thrive.

Having a hiring strategy that supports the development of cumulative experience and knowledge within the team has great advantages. In the early part of the venture, there is significant learning gained about the product, the operations, the technology used, the market, and the customers. This knowledge is important to maintain and build on, as the effectiveness of the team increases. Although having all of your team co-located in one place has advantages, it's no longer critical. With the evolution of digital-working capabilities, many businesses can hire and develop team members from anywhere, which allows for a greater pool to choose from. Culture fit, such as the ability to collaborate, share knowledge, commit to the organization and customer needs, and of course the ability to produce high-quality deliverables, far exceeds the importance of where a person is located. Some positions do require more frequent meetings and face-to-face interactions than others, and closer personal interaction may still be required in those scenarios. But, for many positions today, work can be as effective with fully-remote interaction, or a hybrid model where a 50/50, 60/40, or 80/20 split can be implemented successfully. Therefore, it is advantageous to offer flexibility while hiring in order to gain the needed key resources.

Sales, Marketing, and Administrative Resources

As important as it is to hire skilled technical employees to develop innovative products, it is equally important to have skilled business employees who can develop innovative plans to lead the marketing and sales efforts in the target markets. Here you may require a strategy that includes a combination of internal employees, managed contractors, and marketing platforms that can be used to support the launching of your product. Channel partners, who are established entities that have a strong customer base in target markets, can also be part of the launch plans, although this would require an internal resource to manage the relationship and ensure its effectiveness.

Human resources, licensing, legal, accounting, and other administrative roles are functions that are also required by most businesses. However, I'd only recommended hiring dedicated and full-time resources after the business reaches a certain size in employees or operations. When the business is small, or just starting, these important administrative functions can be outsourced to dedicated servicing companies that can perform them on your behalf in an efficient and cost-effective way. There is no magic number that must be reached to bring those functions in-house, but many companies believe that while the size of the company is below 35 employees, they can be outsourced. Once the company size increases and certain administrative activities require more in-house attention, then you can start considering hiring dedicated employees.

Compensating Your Employees through Direct Pay and Mixed Incentives

A lack of sufficient capital or budget allocation can impact your ability to acquire skilled resources. Therefore, you may need to be creative in your offer to employees based on what is important for them, while satisfying the local labor rules and policies. Consider that some team members may agree to work, initially, for a stake in the company instead of a salary, or they may accept a mix of salary and shares based on meeting certain milestones or success criteria, or by working for a certain time period in the company. To plug gaps, you can also consider hiring freelancers or contractors who have the skills needed and are willing to engage for a period of time or for specific deliverables. If you are considering a partnership model, they may also be able to provide resources, or investment for resources, where this would be mutually beneficial for both businesses.

Ideally, you would raise the necessary funds to acquire the core team needed to launch the business, whether through self-funding or external investments. However, when capital is in short supply, a policy that includes mixed incentives can help your business to access resources and skills needed until sufficient funds become available. It's not unusual for early employees to own some stake in the company; in fact, it is expected by most, as this provides a payoff for the risk taken where the reward can be significant.

For prime employees, who have great impact during the launch, the employee may be given between 0.25% to even a 5% share, depending on their role and contribution. For the later employees, the company could issue an employee stock plan equal to 10% of its total shares, for example, and keep these aside for future employees. The 10% pool would get converted to tens of thousands of shares; new employees could then be offered some shares when hired, perhaps with a condition to meet certain milestones such as staying with the company for a certain number of years or when meeting assigned objectives. When Facebook acquired WhatsApp for $19 billion, it had only 50 to 60 staff members, with more than 450 million subscribers. While the two founders of WhatsApp owned more than 50% of the company, early employees who had a fraction of a percent became overnight millionaires because of the significant valuation of WhatsApp and the conversion of their shares to Facebook stocks. Thus, this can be a very attractive compensation structure and works well for both the company and employees. However, this demonstrates a best-case scenario, and there are many more cases where the company loses its full value and the employees' shares do not pay off. Therefore, this approach may not appeal to everyone. It requires a passionate employee who sees the potential and is willing to invest their time, knowing that there is a risk involved.

Employees also expect benefits that meet the minimum local labor rules and regulations. Vacation time, health insurance, and bonuses are typical expectations. Consulting your human resources (HR) servicing company, or internal HR department, can help you determine the rules you need to comply with and the associated cost. Benefits, and tax obligations, can add up to anywhere from 25% to 40% of their base salary, and sometimes more.

Hourly employees have their own rules as well. You will need to be aware of such expenses as you decide on the pay structure, and the form of relationship with the resources you are hiring.

The leader, the team, and your personal and professional networks are the prime catalyst to translate the business mission and its objectives into a successful reality. Every team member should have a skill set that is relevant and important to their role and to the success of the business in meeting its mission. Your secret sauce is likely residing here.

LEADER'S SUCCESS QUESTIONS

Principle 4: Choose Your Team
Who Will Translate Your Mission into Reality?

Ask yourself the following questions to help you select the team that will translate your mission into a reality:

1. What resources, skills, experience, functions, and roles do you need to launch a successful business?

2. How do you plan to inspire them, to empower them, to reward them, and to develop a collaborative culture to maximize their commitment, effectiveness, and the quality of their deliverables?

3. What key measures do you need to have in place for each role to effectively evaluate each employee's performance?

4. How do you plan to compensate them? Is it pay only, equity only, or a mixture of both? Do you need to offer benefits, and what are the costs of these benefits?

5. What additional resources are needed beyond the core roles? For example, human resources functions and payroll.

6. Is there a need for hiring extra resources from contractors or third-party resources from partners?

Part Three

Transforming Your Business Idea into Reality

Principle 5: Specify Your Product
What Are the Key Functional Components,
the Requirements, and the Long-term Roadmap?

Principle 6: Identify Your Competition
What Are Their Strengths and Weaknesses
and How Do You Compare?

Principle 7: Determine Your Competitive Advantage
What is Your Differentiator and Unique Edge?

Principle 5

Specify Your Product:

What Are the Key Functional Components, the Requirements, and the Long-term Roadmap?

S pecifying your product is one of the most exciting steps when building your company. Whether it is hardware, software, a consumer good, retail, or a business service, defining your product is a critical part of the business. You have developed a long list of great features that you believe will make your product complete, and have created a comprehensive plan of what you want your product or service to offer to your customers, but where do you start with specification?

The key here is to always "plan big but start small." The importance of this approach cannot be overstated when developing your product. The reality is that resources are finite and usually limited, and time to market is highly dependent on what you decide to include in your initial specification. Therefore, prioritizing the "must have" features in your initial product is critical. Decide what features can wait. They could be part of your future roadmap. I can see how a passionate leader may consider *all* features vital to their business and their customers. However, there are some features without which the product cannot function and it will fail to offer the minimum expected value to customers. Those must be in your initial list and part of what is called the Minimum Viable Product (MVP), which we will discuss further in this section.

Starting small and prioritizing is not only important to manage your scarce resources, but also to allow for performing critical validation through testing and optimization of the product in the early cycles of development and before scaling the rollout to wider audiences. This approach will enable the company to be agile, developing features in small, but validated, increments to ensure you stay aligned with the market and customers' needs.

If nature can teach us anything, it's how almost everything in existence originated from a much smaller version of itself before taking its true shape over a period of time. Successful products and services are no exception to this rule. Those who recognize it will deliver the most critical features to customers, and this strong foundation will allow them to move forward in a clear and focused direction.

Articulating the Detailed Functional Components

What do I mean by this? When you're developing a product or a service, it is important to start by articulating it in a language that reflects the business mission and incorporates the knowledge developed through all the research discussed in the earlier steps. This articulation must recognize the goal of solving the "pain points" and address the likely users, i.e., the customer segments that you have determined to serve. The ability to articulate the product and its functionality is the first step in building the solution, and it is an indication of a clear understanding of what the business is going to present to its customers and the value it will provide. Create a requirements document — a detailed functional definition of components — where you describe, with clarity, "what" you are building, and the success metrics that you will use to confirm meeting your goal once the product is created. This document will become a reference for you and your team. Once the main solution is defined at a high level, you then need to work with the product team to detail its major components and break down the features and elements for each. By starting with the big picture to guide your overall desired requirements, it allows you to break the product down into smaller, but cohesive, elements so that it becomes manageable to tackle the "building" stage within a well-defined timeline. This approach enables you to allocate and delegate these smaller components and tasks to individual groups or team members. You should seek help, if you need it, from an experienced product manager who can provide guidance on the process of defining your product and describing it in a way that can be translated to detailed requirements for use in creating it. The clarity in defining your product can help the designers to deliver the initial drafts, help the product and engineering teams to estimate the time and cost required to deliver it, and identify the skilled resources, whether internal or external, including dependencies on partners.

Building a Minimum Viable Product (MVP)

One approach to articulating the functional requirements of a product that has proven to work in many global companies, whether large or small, is the concept of starting with building a minimum viable product (MVP). The MVP approach is a form of

development where the product's core capabilities and initial functionalities are built so that it can be demonstrated to customers, or
tried by early users, to gather feedback. It's an excellent way of
testing core functionalities with friendly groups that might include
yourself, your employees, or a small set of customers, before rolling
it out to the general customer base. This approach not only helps
you to gain insight and optimize the product, but also helps you to
increase capital efficiency before committing larger investment to
scale up production, add more features, and run large operations.
This approach also enables you to build the complete product in
iterations and stages, test it, and correct or add features in each
iteration as you receive market feedback on the product's value.

While the MVP approach is considered a fairly new concept
and is being practiced in high-tech companies, its roots go back
to many decades earlier. The movie, *The Founder*, depicts with a
fair level of accuracy the creation of McDonalds, one of the largest
and most successful fast-food chain restaurants in the world. We
learn that this huge chain, with thousands of locations globally,
started with a single model in a local restaurant in California. The
owners, at that time, recognized the appetite for a fast and efficient food service, as most were slow and frustrating. This was a
pain point that was proven to be true and very profitable to solve.
In order to develop a better model, the owners spent many hours
developing an operational process to perfect the most efficient
preparation of their menu items. Several designs, trials, and actual
tests went into the development of the model. Their mission was
to achieve a faster service than anything else that existed at that
time. The final model was tested in one location: the process of
taking the order, cooking the meal, wrapping it, and then handing
to the customer, was repeated and optimized until perfected.

Only when it received a successful customer response did they begin to replicate the approach in other locations, and later to hundreds of cities worldwide by the new owners. In this way, the company resolved most issues with the process quickly, as discovering and correcting problems in one location is significantly cheaper than correcting the same mistake in many locations.

Today, we see the MVP model follow a similar process. It evolves around defining, designing, and developing the initial core product, then small iterations of testing, optimizing, and increasing capabilities over time. It works well in high-tech development companies, especially with digital products and services where the correction process can be managed through software code updates, and the MVP can be tested quickly by leveraging the availability of online access for most users. While the big picture functionality can be defined early on and according to the business goals and strategy, the detailed MVP functionality is defined by the product manager. The purpose is to produce a working version of the desired product within the shortest possible period of time, typically within a few weeks or months at most, depending on the complexity of the offering. In software development, an agile development approach supports this process, where each iteration can occur within only two to four weeks, referred to as a "sprint." In each of those iterations or sprints, a new functionality can be introduced, an existing functionality can be enhanced, or both. This includes time for testing within each sprint and then presenting to stakeholders and friendly customers who agree to participate in testing and give feedback. Having a set of performance measures to gauge whether you are meeting each of the target milestones will help you to evaluate the success of the product. This approach enables quick and continuous iterations to meet the needs of the business and customers, and the feedback gathered helps to confirm or correct assumptions and verifies customer demand or usability. Once the MVP meets the satisfaction objectives, it can then be rolled out to the larger audience. Additional features and functionalities can then be planned and added in the future, depending on how important they are to customers and according to the roadmap, to increase its value and meet additional market objectives.

Patent Protection

When developing a product, if it is unique and new, you may need to consider filing for patent protection. If granted, it will protect your rights and prevent others from using your own creation without authorization or a licensing agreement. Filing a patent requires a detailed and accurate description of your product in order to protect its design, the materials used, the process, components, or functionality. The downside, of course, is that patents become publicly available and, once filed, even if not granted, they become visible to any entity to review after a certain period. With that said, granted patents can offer great value to your company. Your business may end up with a unique feature in the market that only you can offer, and anyone that wants to use it in their offerings will need to agree with you on the terms and fee to license your patented innovation. This can also raise the value of your company and increase its chances of acquiring investors and customers. There are several rules that dictate the filing of patents, and you need to familiarize yourself with them, or seek help from intellectual property and patenting firms that specialize in assisting companies in the filing process.

As granted patents can offer benefits to your company, so too can they offer benefits to other companies. This means that, while you are working on introducing new innovations, you should ensure you don't accidently infringe on others by checking your product or design does not utilize other unexpired, registered patents. If this does happen, a reasonable change in your product can help you avoid infringement, but sometimes the existing patent may have a big impact on your offering. There are public records that can be searched to verify this. The United States Patent and Trademark Office (USPTO) has tools to enable you to search registered patents and trademarks. Most countries have similar offices or tools for researching registered intellectual property and patents. Similar to patents, trademark protection can help you own the rights for a specific name, brand, and logo design, and stop others from using it once it is registered successfully in your name. If you are using a specific name or logo design for your product, you should verify that it is not registered by another company before you start spending capital and time on marketing. Brand names and logos can accumulate significant value over time, so you should consider protecting them. IBM, McDonalds, BMW, Apple, HP, Samsung, GE, and Polo, to name just a few, are all protected names and trademarks.

Do You Need to Rely on Development Partners and Vendors?

Another consideration when developing and articulating the offering is whether you will be dependent on third party vendors or components in order to complete the delivery of your product. Does your product's value rely on other components outside your business? If so, you'll need to produce an analysis of the availability of those components, and the stability of the vendors or partners that may be needed in order to complete the offering. The quality of the partner and their products, the financial stability of their operations, and the credibility they have in the market can all impact your own business. Therefore, a dependable partner is needed. If a vendor is required in order to complete your offering, then identifying more than one is good practice in case you need to switch vendors for business reasons, such as if one fails to deliver, changes their pricing, or introduces new adversarial terms to the relationship. In such a scenario, you must examine the cost of switching in terms of time, effort, and money, and the impact on customers. I cannot count the number of times I have seen a business or a product launch fail due to the involvement of a less dependable partner, which resulted in the collapse of the whole business opportunity. Similarly, a responsible and dependable partner can be a great catalyst for success. Choose your partners and vendors carefully. Ask if their products or services have been used or deployed somewhere else so that you have a reference point. How financially stable are they? Will they exist for a long time? Do they have liability that may impact the stability of their operations? What is their support model? How fast can they respond to your enquiries, whether it's to fulfill an order, to fix a problem, or to address a new need? What is the payment and money exchange model and terms between you and does that model enable you to operate successfully?

Creating a Roadmap: Responding to Customers, Competitors, and Innovation Opportunities

As important as defining an MVP is to the early stages of launching a product, it is equally important to define a long-term view of the products you intend to introduce in your business. This is typically referred to as a roadmap. The roadmap defines how your business will evolve its offering over the next months and years. A typical roadmap would have a detailed level of functionalities for the first one to three years, depending on the offering type and your company's development cycle, and a high-level description (less detailed) for up to five years and sometimes beyond. For example, a business that is launching an ecommerce online retail store may have a plan to start with offering entertainment electronics such as TVs, game consoles, and audio speaker systems in the first year by accepting payments online through Visa or Mastercard. The roadmap for the second year may include adding financing capabilities, where they would need to modify not only their own software to accept financing applications, but also establish business relationships with banks and other credit-processing institutions for the financing service. Defining this item on the roadmap allows the company to have a head start on this strategic feature and start assessing the resources, time, capital, and partners that may be needed to fulfill these requirements. The product team can also start designing the process and work with external vendors to identify suitable solutions to integrate within their platform. The third year may include expanding their offering to include additional retail categories, such as smart home controls including security alarm systems, smart thermostats, and internet of things (IoT) communication devices. These capabilities will also require advance business and technical planning, possibly six to 12 months ahead of time. Decisions on the features, the customer experience, and the design flow of adding those capabilities need to be made ahead of time and approved by the stakeholders, allowing for research time and a customer validation period. Each of those capabilities may need to be implemented over time as well, and therefore the roadmap will not only include new features, but also enhancements to existing features over time.

The roadmap must be revisited on a frequent basis, to assess any new learnings from the market and customers which might optimize the direction of company, its products, and the priority of the items on the roadmap. Most business environments are in continuous flux, and some industries move faster than others. Pressure from competitors, changes in consumers' demands, and new innovations may mean you need to adjust your roadmap to stay relevant and avoid unnecessary costs. Even at the largest and most advanced technology companies, these drivers of change can push them to act by reprioritizing roadmap items, or introducing new functionalities in order to remain competitive and relevant to its customers. Otherwise, they risk losing their leadership position and associated market share. Samsung and Apple are prime examples of companies competing within a dynamic environment where both continue to adjust their products, based on technological advancement in the hardware and chip industries, new features and functionalities offered by each other, and customer demand for these features. Elements such as screen size, processing power, number of cameras, and battery life are constantly revisited in research and consumer feedback in order to maintain a competitive advantage. They are symbiotic: one company's roadmap is impacted by what the other introduces, especially when the introduced feature receives a great level of demand.

Banks follow the same pattern, whether it is competitive rates on loans, new savings products, or providing a better digital experience. They are always impacted by what customers find useful and important, and by a positive customer response to what their competitors offer.

Developing a business culture that cultivates your team's unique skills is critical to sustaining a competitive roadmap. This means the organization must be built to encourage and reward innovations, and with flexibility to address changes in the business environment in order to adapt when necessary. Ego and bureaucracy can be the greatest enemies for your roadmap adaptability, and for the sustainability of your business. The perfect roadmap and functionality, as viewed by the business, may prove to be less important to customers. Competitors, for example, may introduce features that could upend your roadmap and you will need to be open to throwing out your intended plan and responding urgently by adjusting and reprioritizing your products. The market may surprise you, and competitors may know what should be built before you; it is important to humbly recognize and accept this.

Those who remember the rise and fall of Nokia mobile devices know well how the unique secret sauce behind the company's success was not enough to maintain its leadership in the market. This was mainly due to its relative lack of flexibility in addressing new consumers' demands compared with the competition. A culture of high egos and inflexible execution were behind the fall of a great market leader. They refused to adapt to market changes when smartphones hit the shelves and competitors began to introduce entertaining and useful applications. Nokia's leadership insisted demand was not sufficient to change their strategy and adjust their roadmap. They argued their brand name and the quality of their products were sufficient to maintain their edge over their rising competitors. As we now know, this led to Nokia's downfall in the mobile devices market, a market which has reached close to a trillion dollars as of 2020 and which shows no sign of stagnation. Nokia had a dominating share of 40% or more at its peak in 2007, but that share started to shrink and dropped to less than 3% by 2013 just before it was acquired by Microsoft. By the time they recognized their strategic error, it was too late and they could not recover. Fortunately, learning from their mistakes, they later evolved their strategy, and refocused their efforts to become a relevant player in the telecommunication infrastructure world, supplying cutting edge and innovative wireless network equipment to global mobile operators. However, they certainly lost what was a huge opportunity in the consumer mobile devices market.

With that said, it is important to point out that not all competitors' actions are worthy of emulating. Some functionalities may prove to be less important to customers, and others may fade in value after a short period. Therefore, it is vital to evaluate customer demand and choose the areas where your business needs to stay competitive. The famous saying, "Choose your battles carefully" certainly applies here, as you must allocate your finite investments, resources, and time strategically. While doing so, the business must also be consistent with its mission and objectives and continue to run its operations and performance within its defined culture.

LEADER'S SUCCESS QUESTIONS

Principle 5: Specify Your Product
What Are the Key Functional Components, the Requirements, and the Long-term Roadmap?

Ask yourself the following questions to help define your product, its MVP, and how it will develop (plan big, but start small and agile):

1. What is your solution?

2. How would you define it at a high level, in plain and simple language which any user can understand? This must reflect the research that helped to define the opportunity, the customer pain point that it addresses, and the target customers.

3. What are the detailed functional components of your business? List and describe each in brief, yet with enough detail so they can be clearly understood. Providing clarity on the functional capabilities will help you and your team to translate those requirements into reality.

4. What is the short-term initial functionality that would be necessary in the MVP? How long will it take to develop the MVP? What is the project plan, milestones, and timeline?

5. What is your long-term roadmap? Define the product and functional roadmap for the first six months, and for years

one, two, and three. Outline the mature product view and its evolution after the MVP.

6. What would it take to develop it and offer it in the market? This should outline resources, skills, knowhow, and the availability of components, if there is a dependency on a third party.

7. Are there any dependencies on partners or external entities to complete the offering? In some cases, the more partners or dependencies that are required, the more complex, lengthy to build, and expensive the offering can become. Try to understand the potential partnership challenges ahead of time and be prepared to resolve them as they arrive.

8. What measurements of progress and key indicators are in place for tracking business success?
    a. Define milestones with timelines.
    b. Define customer satisfaction measures.
    c. Define indicators that allow you to identify how the offering is meeting the desired customer value, including customer engagement.
    d. Define indicators to measure how the offering is meeting the defined functional requirements.

9. Would you need to file any patents or trademark protection to reserve the rights to any unique functionality, process, design, logo, or service?

Principle 6

Identify Your Competition:

What Are Their Strengths and Weaknesses and
How Do You Compare?

While focusing on customers and their needs is the most important aspect of the business, understanding your competition is essential to navigate the market and gain a meaningful and sustainable market share. Study and understand who the competitors are in the space you are operating in. Analyze your competitors at industry level, to learn who offers similar products or value. Compare their functionalities and capabilities, including the usability of their products. Understand their go-to-market business model, including where they offer their products, in which markets, and whether they have a physical, or online, presence, or both. What is their pricing in each of the locations they operate in? Develop an understanding of how your business will stack up against theirs. This insight provides guidance on how to optimize your plans in relation to theirs and how your business can compete. Your assessment may also include understanding their financial strength, employee size, the customers they have acquired, what people like most about them, and what people dislike or are dissatisfied with. How fast is your competition responding to customer demand, and where do they struggle in meeting this demand? Furthermore, develop an understanding of what percentage of their business is dependent on the products and services that you both offer.

When assessing these factors, you are trying to identify elements that you may use to develop a competitive advantage and use in your products, your go-to-market strategy, and your marketing messaging. Therefore, it is important to conduct the research with an unbiased perspective and to perform a true readiness review of your own capabilities compared to competitors' strengths and weakness. I discuss SWOT analyses in Principle 7 (Determine Your Competitive Advantage), which is a great tool to use in competitive assessment. One way to check the validity of your analyses is by verifying that you can support your assessment with actual data and not just assumptions or feelings. Present your analysis to a friendly audience, including to internal resources within your company and to trusted advisors, to gain further confirmation. While data may not always be readily available, you may be able to do your own research by reviewing your competitors' websites, social media, search engines such as Google, by shopping at their businesses, going through a buying cycle of their product or service, and talking to their customers. You could also hire a third party to provide insight and assessment to complement your research. Often, when the competition is a known business and has a significant presence, you may find readily-available competitive assessment reports that are developed in the market and offered free of charge or sold for a reasonable fee. In the business world, every competitor who is serious about their product and their customers would continuously conduct these competitive assessments to ensure they maintain a leading edge, or at least to stay within close proximity of the leaders in the industry. Competitors are most likely conducting similar competitive analysis on your business as well. Be sure to do your part.

The following are factors to examine to help you in determining your competitive environment:

1. Who else is offering similar value, products, or services that can compete and gain market share from your business?

Your market share is determined by several factors, one of which is your competitors' size and scale. *How many* companies are competing within your target markets and likely to be offering similar value to your customers is also critical. The stronger the competition is, the harder it is to capture a market share, and most likely your profit margins will suffer too. The automotive industry is a prime example where they invest heavily, not only in the development of their products but also in marketing, to stay relevant. These costs add up, along with price wars in most of its segments, leading to low profit margins that hover around 10% or less. That's where differentiation can help in boosting sales and profitability. For example, innovations in new technologies such as electric vehicles, driverless capabilities, and connected wireless services within the vehicle.

2. What are your competitors' strengths and why are customers buying from them?

It is important to examine what competitors are doing well and understand their strengths in terms of why customers are buying from them. It could be that there aren't other companies offering similar products and therefore your business will change the landscape in this sector once introduced. Or competitors may have a unique value that is expressed in their product's offering, such as competitive functional capabilities, ease of access, brand credibility, and pricing, in which case your entry to market becomes more challenging. Understanding their strengths and why customers buy from them will help you to establish a market-entry strategy to differentiate your product's functionalities, marketing messaging, and your pricing structure.

3. What are your competitors' weaknesses in terms of meeting customers' expectations?

A weakness is typically recognized as being a factor that causes friction in the relationship with the customer. It can cause hesitance in the buying decision and deter customers from doing business with a specific company. The weakness can be in any element of the business, such as in the product's quality, the operational model, trust, processes, ability to quickly fulfill demand, or customer relations. This friction, once identified, can help you to differentiate your business. As discussed in the next principle, it can enable you to gain an edge over the competition and customer acceptance.

4. Which are the top one to three market leaders among competitors and why?

As we examine weaknesses, it is also important to recognize successes among your competitors, especially those with leadership positions in the market. Understanding the reasons for their distinctive success will help you to pin down what your customers care about and how much you need to improve on their offering to achieve market recognition. Prioritizing the importance of competitive features serves you well when it comes to developing your differentiation strategy.

5. Can you project your market share vs. your competitors? Will that share be sufficient to cover your operations and desired profit margins?

Be ambitious, but reasonable at the same time. It is rare that you will have 50% of the market, for example. In some industries, a 3-5% market share is considered a leader. We will cover market share projections and calculations in Principle 8 (Know Your Financials).

In the next principle, Developing Your Competitive Advantage, I will discuss SWOT analysis. This is an important tool that helps you determine how your product stacks up against competitors, as well as opportunity and market threats. Therefore, developing a good appreciation of your competitors' position in the market is critical for this next stage and can be strategic once leveraged within your product differentiation and market entry plans.

LEADER'S SUCCESS QUESTIONS

Principle 6: Identify Your Competition
What Are Their Strengths and Weaknesses and How Do You Compare?

Ask yourself the following questions to help articulate and understand the competitive landscape:

1. Who else is offering similar value, products, or services that can compete and gain market share from your business?

2. What are your competitors' strengths and why are customers buying from them?

3. What are your competitors' weaknesses in terms of meeting customers' expectations?

4. Which are the top one to three market leaders among competitors and why?

5. Can you project your market share vs. your competitors? Will that share be sufficient to cover your operations and desired profit margins?

Principle 7

Developing Your Competitive Advantage:

What is Your Differentiator and Unique Edge?

Now that you've identified the competition, and evaluated their strengths and weaknesses, you need to define your own offering's competitive advantage. What differentiates your product, service, solution, or business overall from others in the market? In other words, if your offering launched today, how would you stack up against competitors and what would compel a customer to buy yours over other existing options? How are you going to compete and win a meaningful share that can meet your organizational goals?!

Your competitive advantage can be a tangible advantage or an intangible advantage that is not visible to competitors. A tangible differentiation can be achieved through a unique or improved functionality that can meet customer needs better than what already exists. This might be improved usability, ease of access, a differentiated quality of your product, or a better pricing model. For example, if research indicated that battery-charging speed was one of the most important features to customers when buying an electric car, then it might be that your electric battery provides the fastest charging speed in the market. The key is understanding how your offering, and its unique characteristics, meet your customer's demands more than the offerings of others in the market.

An intangible advantage, on the other hand, is more difficult for someone to copy. It could be the brand or your projected image, which may reflect quality, reliability, luxury, great customer service, or friendliness. A simple example is when an identical product is offered by two organizations at the same price. A customer may choose to buy from one over the other based on their assessment of the company, such as the credibility it has in the market, brand awareness, the support model, the ease of contacting and communicating with the business, and better customer service. In fact, many customers are willing to pay more for the same product if the business has better market credibility and dependability. For example, when buying a laptop computer, multiple online retailers may offer the same computer with different prices. However, the user may choose to buy from reputable retailers such as Best Buy, Costco, or Walmart rather than less well-known retailers, even if it's available at a lower price, because the credible retailer is more likely to exchange or return the product if there is a problem. Therefore, price alone cannot always be a differentiator. Quality of service, confidence, and trust in the business can far exceed the value of discounts, even in price-sensitive segments. I don't intend to suggest here that only large or well-established institutions are capable of developing a credible image with customers; rather, I want to ensure that you pay attention to the intangibles, such as trust and credibility, and make them part of your organizational goals.

There are several competitive assessment models you may

want use to help you develop an understanding of your unique edge, and what more is needed to strengthen your competitive position. There are two in particular that I'd like to highlight:

1. SWOT Analysis: The first is what is widely known as Strengths, Weaknesses, Opportunity, and Threats (SWOT) analysis. This type of competitive analysis can help you evaluate and understand the competitive position in the market and how you stack up against competitors. This insight, when coupled with the analysis discussed in Principle 1 (Define Your Opportunity), can help you to determine where you can enhance your competitive advantage.

a. Strengths

What unique strengths and advantages does your business has over its competitors? This could include product capabilities, relationships, market knowledge, online and physical presence, skills, pricing structure, patents, and any other strength that can help your business to differentiate itself.

b. Weaknesses

What are the weaknesses of your business? Don't be afraid to list what you believe you are lacking, especially when compared with the competition. This will help you to understand the challenges you may face and plan to overcome them, either by further differentiating yourself, or by mitigating those weaknesses over time.

c. Opportunities

What are the opportunities that you can leverage immediately, and then in the long-term as you scale? This can help you focus on immediate action to support your business in securing your first set of customers and prioritizing the products to support them.

d. Threats

What are the threats to your business in terms of emerging competitors, declining demand, and other risks as I discuss in Principle 12 (Evaluate Your Risk)?

SWOT Analysis Template

Strength	Weaknesses	Opportunity	Threats
1.	1.	1.	1.
2.	2.	2.	2.
3.	3.	3.	3.
4.	4.	4.	4.

2. Porter's Five Forces: Another useful analysis technique is to evaluate and determine your power level in the value chain, and the environments you are going to operate within, using Porter's Five Forces analysis. This was developed by Michael E. Porter, a Harvard professor. His assessment proved to be highly successful in identifying factors that can impact the profitability of a business.

Porter's Five Forces are:

a. The bargaining power of the customer or buyer. When your customers have bargaining power, such as having many options to choose from, this can influence your price, and the product's functionality and quality. The customer can demand to receive more for less.

b. The bargaining power of suppliers or vendors. When suppliers have bargaining power, such as scarcity of competitors or if it will cost significant capital for you to replace them, they can force you to pay more, and can influence the quality of your product if their standards are low. During the COVID-19 pandemic, for example, sellers of protective gear have raised their prices manyfold due to scarcity and rise in demand.

c. The threat of a substitute. This is the risk of your product being replaced if other products become available that can offer similar value at a lower price. If the switching cost for the customer is low, then they can easily leave you for others.

d. The threat of new entry. If your business has few barriers to entry, then new companies can easily establish similar products to yours and enter the market, which will reduce your market share and likely your prices as well.

e. The threat of current competition. Entering a market that has many existing competitors will increase your costs and reduce your profit, as you will need to compete more aggressively in the market.

Porter's Five Factors

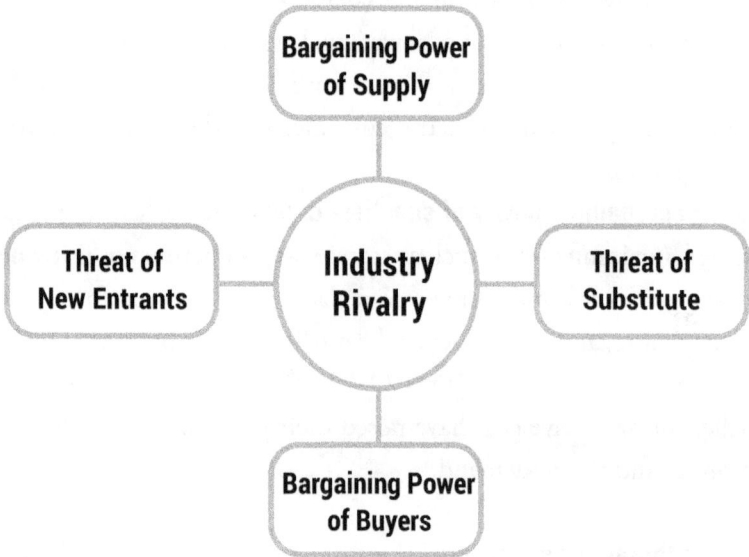

When creating and developing your differentiation, you need to define how unique your competitive advantage is and how difficult or easy it would be for others to replicate it. One of the ways to protect your unique edge, if it's dependent on design and functionality, is through filing patents. A protected patent can certainly be a strong differentiator, as others will not be allowed to offer the same product to market during the protection period, even if it is possible or easy to replicate. However, if your differentiation is intangible then it's more difficult for competitors to observe and copy. A superior customer service, for example, which yields customer loyalty and greater engagement, can come from the culture the company develops and is often driven by internal processes that are not visible to competitors.

LEADER'S SUCCESS QUESTIONS

Principle 7: Determine Your Competitive Advantage
What is Your Differentiator and Unique Edge?

Ask yourself the following questions to discover your competitive advantage:

1. Why would customers use your product or service instead of others?

2. What makes your offering distinctively different, and how valuable is this to the customer?

3. How are you achieving your competitive advantage? Is it tangible through product features, pricing, distribution, time to market, first market mover etc.? Or is it intangible, such as better customer service, a better engagement process, relationships and connections, brand awareness, easier access, a better experience, or free or faster delivery?

4. How does your business stack up against others when performing SWOT analysis? Are you counting on gaining market share from competitors? If so, describe how you will develop a more competitive operating model.

97

5. Can others easily copy your competitive advantage? How difficult is it for others to acquire the same differentiation? Do you have patent protection granted to your product, design, or process?

6. Do you, or your team, have unique experiences, contacts, and skills that can provide an intangible competitive advantage? Can they support the differentiation of your offering in the market? Your relationships with potential customers, your knowledge in the industry, and your skills in developing a solution can all help give you a unique edge.

Part Four

The Strategy for Making Money

Principle 8: Know Your Financials
What Are Your Cost, Pricing, and Revenue Structures?

Principle 9: Build Your Go-To-Market Strategy
What Will Your Initial and Full Market Rollout
Look Like?

Principle 10: Secure Your Funding
What Capital Do You Need and How Will You
Find Investors?

Principle 11: Prepare Your Pitch
How Can You Convince Investors, Partners,
and Customers to Get on Board?

Principle 12: Evaluate Your Risk
What Is the Likelihood of Failure and How Will You
Prepare for and Mitigate Your Business Risk?

Principle 8

Know Your Financials:
What Are Your Cost, Pricing, and Revenue Structures?

The stability of your business is dependent on the health of your financial operations. One of the common characteristics that you will find with successful business leaders is that they know their numbers well. Understanding business finances is essential for sustainable longevity. Know your costs, your pricing, and your revenues. Only then will you be able to determine what needs to change and when to maintain a healthy bottom line and increase the value of your business over time.

A leader doesn't require the accountant's level of knowledge or the chief financial officer's skills but, put simply, they must understand the basic business finances in order to make sound decisions. At the beginning, a business should develop a good understanding of what it would cost to launch the product, operational expenses, and the likely income required to cover its operational costs after launch and produce some profit to support its growth. If a business intends to be profitable and provide value to its stakeholders, then projecting what that breakeven point is, and the time it will take to reach it, is important. For non-profits, understanding your costs and spending rate allows you to define the income needed to cover these activities and sustain operations.

Understand Your Costs

In any business, there are going to be initial setup costs and then an ongoing operational cost to run it. It is important to determine what the initial cost will be to launch the business, the product or service, and how much is needed by when. The initial cost may include your own time and, if you are expecting to be fully dedicated to the business, your own living costs, whether that's through the business, through another source of income, through borrowed capital, or through your own savings. You will also need to cover, as applicable, the cost of some of the early research work, equipment, devices, supplies, licensing, legal incorporation costs, office space, contractors, initial design work, consultants, and initial implementations such as producing a demonstratable version of the business. Not all businesses will require all of these costs, but you need to make sure you account for those that applies to yours.

In addition to the initial setup cost, there will also be a set of recurring fixed costs and variable costs that will need to be covered on an ongoing basis to produce the product or service and sell it to the market. Fixed costs are those that do not change often and must be paid in every specific period, regardless of the status or the income of the business. They can include rent, payroll, payroll taxes, insurance etc. There are also variable costs, which are those that typically change based on the scale of the business, such as sales cost, and operational cost where materials, contractors, and equipment may be needed. There may also be a need for hosting of software applications in the cloud where the more usage there is, the more processing power and data capacity is needed. Such services will increase the business's cost as they occur.

Knowledge of costs, and factors that can increase or decrease them over time, is required to project the initial funds needed to launch and run the business. This will feed into developing the pricing models and revenue projections. It also provides you with insight into how much financial flexibility you might have in offering discount models, or ease of payments to gain customer traction. Indeed, understanding your costs can help the business to differentiate its offering in the marketplace through offering unique pricing models that acknowledge the short- and long-term cost of the operations. For example, you could gain economies of scale in data-hosting costs where the growth in usage does not increase linearly, but rather grows at a slower rate than the increase in number of new customers, data usage, and revenue.

LEADER'S SUCCESS QUESTIONS

Principle 8: Know Your Financials (Cost)
What is Your Cost Structure for Launching and Operating Your Company?

Ask yourself the following questions to construct an understanding of your cost models:

1. What are the total initial funds required to launch the business, the product, or the service? Your set-up costs could include legal costs, incorporation and government registration, licensing and other administrative costs, lease or real estate expenses, equipment, supplies, and furniture.

2. What taxes, interest, and other financial obligations do you have that will impact sales, revenue, and cost projections?

3. Do you plan to pay yourself or cover your living expenses? If you don't have other sources of income, how will you cover your personal obligations?

4. What is the cost of developing the MVP?

5. How much will it cost to test the MVP in the market with customers? Would trials be needed?

6. What are your fixed costs and your variable costs?

7. As applicable, what is the cost of producing a single unit of the product?

8. How can you improve your operational efficiencies over time without compromising the value or quality of your product?

9. How many resources does the business need in order to produce the product? Are they full-time, contractors, or a mix?

10. How long will it take for the business to be revenue ready and sell the first product?

11. How much would it cost to become customer ready, i.e., to engage and sell to customers?

12. What other costs are needed for market readiness? Is there a third-party cost that needs to be considered?

13. What are the marketing costs needed to bring awareness to market?

14. What are the sales costs? How much would it cost to acquire customers and sell the offering (resources, advertising, tools, etc.)?

15. Have you defined the cost to fund the initial launch and then the roadmap over the next one to five years? Segment the cost per milestone that needs to be reached over a certain timeline.

Define Your Pricing Model

When pricing your offering, you are bound by several factors: the cost to produce the product or service; your competitors' pricing; your competitive advantage and how differentiated it is; the perceived value by the customer; the customer's demand of the product; and the availability of similar products in the market. All these are factors that determine your pricing model or the structure by which you may offer your products or service in the market. There are several pricing structures that you can use to offer your product to market, and three pricing models that are helpful in determining your price value: cost-plus, value-based, and market-based models.

Pricing Structures

In terms of structure, your pricing may be offered in different forms, depending on your business and the types of products you offer within a given industry. Although there are many forms and structures which are discussed in this book, I'd like to highlight the seven pricing structures below, as they are applicable to the majority of businesses.

Buy-to-Own

In this model, a customer pays directly for the service or product they buy. It is typically a one-time payment to have the perpetual rights to own the product, such as when buying a home, a paint, or a computer. It's the same with businesses that acquire equipment, tools, and supplies. An amount is paid directly by the business to the vendor or supplier for the perpetual ownership rights.

Lease or Subscribe-to-Use

This is when a customer pays for gaining a temporary and conditional access to your service or product over a period of time, such as a monthly or annual basis. For example, when you lease a vehicle, rent a home, or gain access to a streaming service such as Netflix. The monthly payment covers the cost of using the product during this period. Often, an annual subscription rate is offered at a discount as it ensures a longer-term commitment by the customer. The discount amount is bounded by your financial objectives, your costs, and research around your customers' responses to a specific discount rate.

Third-Party Sponsored

This is where the user acquires a product or service and benefits from it for free. It does not cost him or her any fees. Rather, a third party that has interest in the customer using it subsidizes or pays for it. TV network broadcast channels have used this model for many years and still do. Viewers access the channels over the air for free, and this is supported by revenues collected from commercials. Google Search followed this model and pioneered it in the digital advertising space. Those who use the search engine do not pay for the service, but advertisers who appear within the content do, as discussed in the pay-per-use structure below. Facebook, and the free version of Spotify, are also examples of this structure as their business customers pay to place ads within free content that the public engage with. Many government-supported services used by the public also fall into this pricing category. For example, emergency alerts, E911, the PBS channel, and other educational programs, are services used by the public for free. The companies get paid for producing these products by the sponsoring government entity as it has an interest in the public benefiting from these services.

Pay-Per-Use

This model enables customers to gain access to a specific service only when used, based on certain measures. Businesses who buy advertising on Google Search, for example, can sign up to the service and decide how much to spend on a daily or monthly basis. Some companies decide to spend a few hundred dollars a month, while others can be spending tens of thousands based on their market goals and budget. This model takes multiple forms in digital media, which includes pay-per-click (PPC), where the business customer pays only when a user clicks on their content, pay-per-view (PPV), where the customer pays per displayed impression of their promotions within personalized content, and pay-per-action (PPA), where a specific action is tracked, such as when a user fills in a survey, buys a product, registers for an event, etc.

Base Subscription with Pay-as-You-Go

Here the customer pays to gain access to a service with a base monthly fee, then pays per use based on how much they use the service. An example would be hosting your application using a cloud service provider, such as Amazon's AWS. Amazon charges a low basic monthly fee to have a customer connect and gain access to a set of hosted services and then the customer pays more based on the services they use and the size of such usage. The more power, capacity, and data transactions a company uses, the higher the monthly invoice will be. This is an advantageous model for businesses, since they pay only for what they use and when they use it. They can optimize their spending as their needs change from one month to another. Obviously, Amazon has to continuously plan and monitor its own platform to project future usage and be prepared for the variations in utilization and growth rate from so many customers.

Freemium

This model allows a given company to offer their products or services in two or more classes, where the base class can be offered at an introductory level for free and, once acquired, the customer is offered additional services and upgrades with premium capabilities for a fee. Zoom, for example, the video conferencing and collaboration application, has chosen to offer a free entry-level service where the account holder can register and start hosting 45-minute sessions for free, for up to 100 attendees. However, if the user needs longer meetings for more attendees, then you have the option to buy an annual license. LinkedIn, the professional social network, is also a good example of this model, where anyone can create their own profile for free. But there are premium services, for a fee, allowing sales executives to identify potential buyers, and recruiters to identify professional candidates. Members can also buy premium access to services to help them find job opportunities, such as seeing who has recently viewed their profile. Many public events or conferences also follow this model, where they offer free general access to an event and then offer specific key activities, such as presentations by well-known speakers, for a fee.

Buy-to-Own with an Annual Subscription Fee

In this model, the customer pays for perpetual rights to acquire a product or service and then, to keep the service running, they pay additional fees on a periodic basis. Many business-to-business products use this model. For example, a business may buy some office equipment, like a photocopier or a billing software application, and then pay annually for maintenance and updates. Many consumer products also fall into this model. For example, when buying a home security system, the customer pays for the equipment to be installed at a desired location, but then pays a monthly or annual subscription fee for monitoring of that location 24/7 by a dedicated security facility.

Pricing Models

Whether you are pricing your product directly to the end user, subsidized by another business, or using a freemium model, your pricing determination formula will have to use one of three pricing models below.

Cost-plus Pricing

Some businesses follow the simple cost-plus model, where you calculate the fixed and variable cost to produce and sell the product, and then add the desired profit margin. However, there are downsides to this model. While it is obvious that your price needs to be higher than the operational cost of producing the product, customers in general do not care about your costs. They will compare your price to others and, if it's significantly higher than competitors, they will be looking for ways to justify the price differentiation, such as superiority in quality, in service, in trust, in safety, in ease of buying, etc. If they cannot justify the increase, they will not buy.

It's important, when using the cost-plus model, not to base pricing on a small volume of projected sales. While variable costs grow with sales, fixed costs do not, so in order to cover all costs including fixed costs, the price per unit may end up higher for a small number of units and much smaller for a larger number of units. Therefore, a business can price itself out of the market if they're determined to cover all their fixed costs based on low projected sales. For large operations, the fixed costs end up being a very small part of the overall cost and therefore have less impact on pricing. Careful analysis of the percentage of the overall fixed cost per sold product is needed in order to avoid overpricing. For example, it may cost you $10 per unit to produce a keyless RFID door card, with fixed monthly costs of $10,000 for rent, labor, tax, etc. If you project sales of only 100 cards in a given month, then you will need to price the card at a minimum of $110 to breakeven, so that the total revenue of $11,000 can cover both the cost of making the 100 cards and the fixed business costs. However, if you are projecting sales of 1,000 cards a month, then you would only need to price the card at $20 in order to breakeven and cover both the material cost and the fixed cost, which in this case would come to $20,000. Of course, to realize a profit, you would need to add an additional margin to the $20. If you added $5 to the price that would result in a net profit of 20% ($5/$25=20%), and, in this case, it will produce a $5,000 profit ($25,000 sales – $20,000 costs).

Therefore, I recommend, when using the cost-plus model, that you start pricing by using the variable cost, and then add a margin to cover the fixed cost and the additional profit that you desire while staying competitive in market. Some competitors may price their product under its cost as they may be using it as a "loss leader." This is a technique to lure customers in, so they buy other products from the company that are more profitable. When this strategy is used by your competition, it can undermine your business if it's highly dependent on that product. Therefore, while cost-plus pricing is a valid model to use, it is not favored by many startups and, instead, they tend to use value-based pricing as I describe next.

Value-based Pricing

As I discussed earlier, customers don't care about your business costs, nor do they want or need to see them; they only care about the value of the product and whether they can justify its pricing and can afford it. When competition is not forcing the price into a specific range, and your product is unique, you can choose to price it based on the perceived value by the customer. While the price obviously still needs to include a margin above the cost of making it or offering it, to ensure it covers variable and fixed costs and turns a profit, that margin can be determined based on your customers' willingness to pay and the demand for the product at a given price point. When introducing value-based pricing, you may want to perform a customer study to determine the optimum price point at which to launch your offering in the market. You may then adjust your pricing over time to meet the desired sales targets and align with your customers' willingness to buy. For example, you may find a user is willing to buy an iPhone or a Samsung that is multiple times more expensive than another device simply because of the high perceived value by the user, as expressed in the quality, trust, product features, brand, and other personal values being satisfied. The user does not care how much the phone cost to make. Differentiation can drive the value of the product upward and therefore companies can demand a premium for their products, irrespective of the costs of production. Value-based pricing is a favored model, especially for digital products, where the material and fixed costs to reproduce are small, yet the price can remain consistent with each new sale made. For business sales, value-based pricing is also popular. Think of an invention or a service that can reduce the cost for a mobile operator or an electric grid distribution company by tens or hundreds of millions of dollars. In this case, valuing your offering anywhere between 5% to 25% of the savings in cost to the vendor is not an unreasonable price. Similarly, your offering may be helping your clients to generate a significant amount of new revenue, and you can price your offering based on this added value. Obviously, there can be other costs associated with each sale, such as customization or integration, but, in general, this model can be more financially advantageous for innovative offerings.

Market-based Pricing

When products are offered in a very competitive environment, where similar offerings or substitutes can be purchased easily instead of your product, the market will determine your price. This often means that the product cannot be differentiated sufficiently and is not unique enough, and therefore you must be competitively priced. However, as discussed earlier, it is never good practice to compete by price alone – you need other differentiators to gain competitive advantage. This may drive your business to fail, especially if your competitors have multiple profitable offerings in the market and can thus make their substitute a "loss leader" to drive down prices, reduce your margins, and drive you out of business. Market-based pricing does not mean you must price your product to match your competitors exactly. You can price it higher or lower, but you will need to keep it within close proximity, unless the higher price can be justified by your competitive advantage.

The chosen pricing model and its structure can be the key for business success. One product may find great success with a particular model but lose favor with another. Research your customers, their typical purchasing behaviors, and your competition, and truly understand the value of your product to determine not only the pricing but also the model that will ensure sustainable engagement with your product, your business, and your brand.

LEADER'S SUCCESS QUESTIONS

Principle 8: Know Your Financials (Pricing)
What is Your Pricing Structure?

Ask yourself the following questions to determine your pricing model:

1. What is the pricing structure that is suitable for your business and who pays for the product?
   a. Buy-to-own.
   b. Lease or subscribe-to-use.
   c. Third-party sponsored – when a product for consumers is subsidized or paid for by a business customer or governments e.g., advertisers on Google Search enabling the public to use this service for free.
   d. Pay-per-use.
   e. Base subscription with pay-as-you-go.
   f. Freemium - where the basic service is free with premium subscription options or upgrades for a fee.
   g. Buy-to-own with an annual subscription fee - such as an initial purchase fee, then a monthly subscription for annual support and maintenance.

2. What is the most appropriate pricing model for determining the price value of your offering? Is it cost-plus, value-based, or market-based?

3. How are your competitors pricing similar products?

Calculate Your Revenue Projections

Projecting your revenue helps you assess the future health of your business. The confidence level in the generated estimate is highly dependent on the underlying assumptions, which must be well researched, realistic, and not developed out of thin air. There is often a difference between the desired revenue and the likely revenue that you will realize. A company's revenue is the total received income in a given period as a result of selling the products or services offered by the business. While calculating past revenue is factual and can be easily tracked and audited, projecting future revenue is more difficult to accurately calculate. It is typically an estimate that is driven by your company's past performance, if there is any history, your company's financial goals, your go-to-market strategy, and your confidence level to capture or generate needed demand within each target market.

Some companies use highly ambitious growth rates as a target, such as 10X (achieving 10 times the growth rate over last year or last month), 100X, or 1000X. With that said, when developing a revenue projection model for a new product, it must be supported by an executable plan reflecting a solid understanding of the market size, competition, differentiated advantage, customer demand, pricing, and the overall go-to-market strategy. You may project significant growth, but be sure you can defend this with believable assumptions. I certainly encourage any leader to challenge themselves and their organization to aim high, as this drives effort, team collaboration, and innovation. You just have to look at the myriad business success stories to see that we have always found ways to turn the unrealistic into reality: the aerospace industry that started with a simple, single-propeller plane and now produces engines that can reach space and land on other planets; the transformation in wireless communication, with its extreme growth in data speed and capacity that is supporting new economies across the globe; the growth in the medical field, from antibiotic medicines to preventive vaccines and the high-tech diagnostic devices that can find abnormalities in seconds and equipment that can allow non-invasive treatment; all these innovations and many more, including renewable energy, electric vehicles that drive themselves, and the internet which has democratized knowledge and made it accessible to anyone. These hypergrowth innovations are not slowing down. On the contrary, they will continue to experience unbelievable growth, year after year, at an exponential rate. They've produced similarly remarkable growth in revenue for the companies that are leading those industries. When a slowdown occurs in an industry, it is still typically due to another astonishing technological advancement creating market disruption and a shift in demand.

Revenue Projection Models

Projections can follow two approaches:
1. A top-down method based on the overall market size with a reasonable assumption of a likely captured share of the market.
2. A bottom-up method using projections of likely sales on a weekly or monthly basis, based on the actual go-to-market and sales plan.

The bottom-up method can often produce a more accurate projection and expectation and is useful to set sales goals and quotas in the organization. The top-down method, on the other hand, can help determine the scale of the opportunity and its potential, which is useful in the early part of business analysis.

The Top-Down Revenue Projection Method

The top-down method, as the name indicates, is a high-level market share assessment, based on the total estimated market size. It uses the below formula:

Assume the market size of a given industry is $10 billion. A company estimates that they will gain a 3% market share. Therefore, their projected revenue is $300 million.

One may challenge the assumption by asking, "What is the basis for assuming a 3% market share? Why not assume 15%, or 1%, or less?"

When using the top-down method, many people fall into the trap of using it to project the likely revenue in their first few months, or even years, when they are unlikely to immediately take market share from their established competitors. This usually ends up overstating and inflating the projection and misleading the team, stakeholders, and investors, especially if it's not supported by facts and data.

While this method has inaccuracies, it can help you develop an understanding of the potential scale of the business. You can look at what a large market share would look like, as well as a medium share, and a low share compared to your competitors. This should help you make a decision about what market share position should be targeted and by when, including the development of an actionable plan to reach it. This leads us to the bottom-up projection.

The Bottom-Up Revenue Projection Method

This model uses the facts and figures from the business' projections of sales using reasonable and supported assumptions. Revenue projections need to be as realistic as possible and tied to the strategic go-to-market plan. This includes assumptions of sales volume month by month, market penetration rate, the growth rate month over month, and the competitiveness of the offering. The marketing plan, the geography and online presence, pricing, competition, sales resources, and how closely the offering meets customer demand, including the ability to fulfill demand and the overall speed of rollout, are all factors that need to be considered when developing the revenue projections.

Ideally, the projection is based on estimating the sales, month by month, for the first one to three years. This allows for developing a more accurate bottom-up projection using reasonable assumptions on the penetration rate and the determined pricing.

If you are targeting business customers, then you will have a defined segment of clients which will be easier to pin down and calculate through research. You should use this initial set of likely customers and outline expected revenue from each and by when. If the customers are consumers, then you are working with a much wider pool of potential buyers and segments. You will need to reasonably project how many units may be sold on a daily, weekly, and monthly basis with a growth curve that is driven by the implementation of your go-to-market strategy and sales plans. For the revenue projections to be believable, develop a model that increases in units sold per short intervals such as daily, weekly, or monthly, and not a single, annual number. For example, if your business is selling a video-streaming subscription, your growth rate within a target market may look like 70 new customers in the first week, then 250 in the second week, 350 in the third week, and so on, depending on the number of sales associates, marketing and advertising budget, and the speed at which you can onboard and support those new customers. The fact that you may end up with 1,000 new customers per week by the end of year does not mean you can assume the same for the first month in your projection. You may also need to account for returns; cancelled subscriptions; cyclical variation based on the season, holidays, and pressure from your competition who may increase their marketing and advertising or structure new competing services as a response to your offering, which all can impact your actual growth rate and projected revenues.

You must ask yourself what it would take to hit those projected revenue numbers based on the estimated number of customers you need to acquire, the determined price per unit, and the assumed growth rate. Similarly, for business-to-business, you must make assumptions on the number of business customers you will be able to acquire based on your knowledge of the expected sales cycle, which can be weeks to months, and the size of transactional business to be generated with each of those customers.

Given the natural uncertainty in the projections, it is reasonable to develop three different models. The first should reflect an ambitiously high projection that represents the best-case scenario, the second a middle-level view that may reflect the most likely scenario, and, thirdly, a more conservative view taking into account higher levels of risk that may impact your operations and therefore your revenues. These three models can offer insights into how the projections of revenue change with variations in the sales volume, which can also support making investment decisions around the sales and marketing budgets.

Below is an example of cost and revenue projections for a company that sells mobile data. As every business can be different from another based on stage, maturity, location, and industry, resources and cost may also vary significantly. Salaries and costs in one company may vary by up to 10 times the example below, in either direction. Furthermore, the skills of the founders may have an impact. For example, some may perform much of the listed services (initially) by themselves, thus reducing the estimated cost considerably, while others may be more focused on their own specific roles and require hiring the dedicated resources to achieve their goals. Therefore, please use the examples below for demonstration purposes only. I've seen founders who pay close to nothing to themselves, and others who pay significantly higher. There is no one answer for those figures. They are set initially at a certain level, then deviate considerably based on the success rate, the growth of the company, and the maturity of the business.

Table1 below represents an example of a monthly estimated cost for launching a new startup or a new product within an organization. Businesses may use different role titles, and the functionalities may be carried out by different resources based on the structure of the organization. In the example below:

■ Chief Executive Officer (CEO): Sets and leads the company's direction, vision, strategy, and overall execution plan to meet corporate goals. For startups, the CEO may also fully perform, or share duties for, any of the additional functionalities, including product management, development, and sales and marketing. The CEO would also own, influence, and approve the product's roadmap, market strategy and GTM, market and operators' global relations, and overall operations and accountability.

121

Over time, the CEO would need to pull back and focus more on the executive direction of the company to meet its financial and customer growth goals and allow dedicated executive and non-executive resources to have full ownership of their roles.

■ Chief Marketing Officer (CMO): Owns sales and marketing leadership, including go-to-market plans and execution, distribution channels, and market partnerships. They also have responsibility for providing market information that fuels increased innovation, differentiation, and accountability for customer growth, and customer support, relations, and management.

■ Chief Product Officer (CPO) or VP of Product Management: This is a role that sometimes is shared with, or performed by, the CEO, CMO, or CTO in smaller organizations, as the founders tend to lead the product definition and design; but it becomes more pronounced and detailed in more mature organizations which allocate a dedicated lead resource, or resources, to it. It may also take the form of a vice president of product management and innovation. The product leadership position sets the requirements and leads on defining the product details including design, functionality, usability, differentiation, competitive advantage, and collaboration with all other involved organizations including sales, finance, legal, engineering, technology, and support. The product leader may partner with the CMO and CEO to define and lead the go-to-market strategy and define complete rollout product plans, including financial and customer goals, and pricing strategies. It is also possible for the CPO to own the Profit and Loss (P&L) management and ensure that market targets are met.

■ Chief Technical Officer (CTO): Leads the engineering and technical development teams and organization. They focus on efficient and innovative delivery of the specified product requirements.

The technical leadership, along with the product and marketing leadership, stay attuned to the latest technological advancements, developments, and innovations in the associated industry in order to maintain a competitive edge in the development methods and functionality of the offering.

■ Chief Financial Officer (CFO): In a mature organization, this is a key role to lead the capital management and financial performance of the company, including funding, leverage, debt, expenses, and revenue forecasting (including tracking the financial health of the organization and required adjustments). In smaller organizations, this role is typically shared between the CEO and the other executive team members.

Head Count	Budget Item Definition	Month 1	Month 2	Month 3	Month 4	Month 5	Month 6	Total
								In US$ 1000s
1	CEO and Chairman	9	9	9	9	9	9	54
1	CMO: Sales and Marketing Leadership	9	9	9	9	9	9	54
1	CTO: Overall Technical and Technology Leadership	9	9	9	9	9	9	54
	Add CPO, CIO, CFO, VP of Product and Innovation, and other roles as per the business stage and needs							
Initial 2 to 3, or more according to need	Technical Resources: developers/ testers/system engineer/ project management/ data analytics	8	16	24	24	24	24	120
6	Total Internal Full Time Head Count (FTC) Direct Cost	35	43	51	51	51	51	282
	25% Tax and Overhead Cost (actual Percent can vary to 35%, 45% or more based on actual employment taxes, benefits, and other overhead cost)	8.75	8.75	8.75	8.75	8.75	8.75	52.5
	Total	43.75	51.75	59.75	59.75	59.75	59.75	334.5

External HC (Contractors)									
0.5	Part-time Contractor: CFO-Accounting/ Auditing/Internal Controller – Outsourced initially in this example, or shared within an organization	2	2	2	2	2	2	12	
3 to 4	Full Time Contractors - Technical Development (offshore in this example)	10	10	10	10	10	50		
1	Contract to Hire - Sales and Marketing Support to work with CMO. Depending on business type and its dependency on sales staff, the initial count can increase, and may be a full-time employee	8	8	8	8	8	40		
1	Contract to Hire - Company General Administration and Office Operations, Offshore Contractors Management, HR, Administration, Procurement, Billing and Collection. Can be outsourced initially to different servicing companies as well	2	4	4	5	5	5	23	

2.5 Expenses							
Total	4	24	24	25	25	25	127
Legal Services, Payroll Services, External Auditor	3	3	3	3	3	3	18
Marketing Expenses	3	3	3	3	4	4	20
Travel	15	15	15	15	15	15	90
Rent (to accommodate flexible shared work space)	2	2	2	2	2	2	12
Supplies/office equipment/furniture	0.5	0.5	0.5	0.5	0.5	0.5	3
Bills-Fixed	1	1	1	1	1	1	6
Can also account for building a demo, engaging in friendly trials with customers, and performing focus group research							
Computers/ Servers	10	5	5	0	0	0	20
Software	4	4	2	2	2	2	16
Miscellaneous - Contingencies	5	5	5	5	5	5	30
	43.5	38.5	36.5	31.5	32.5	32.5	215
Monthly Total	91.25	114.25	120.25	116.25	117.25	117.25	676.5

Table 1 – Example of Monthly Operational Cost

Table 2 below represents five years of forecasted costs. Beyond the first two to three years, the forecast is highly variable and has a much lower confidence level. However, it is used to provide a high-level view of future capital needs based on planned expansions in product roadmap, markets, operational support, and a more mature organizational structure where the number of employees, salaries, and benefits increase as financial goals are successfully met. In the below example, the five-year estimated operational cost, or capital needed to run the business, is summarized below. The figures were arrived at by using Table 1 and increasing the roles and cost items per year based on the target growth goals.

	Yr1	Yr2	Yr3	Yr4	Yr5
Company Operational Cost (Expense/ Capital)	$1,353,000	$4,958,000	$7,770,000	$15,540,000	$31,080,000

Table 2 – Represent the five years forecast of operational cost

Table 3 below represents a revenue projection structure. The "penetration rate," which is a major variable in this revenue equation, can be calculated in two ways. One is a general, high-level approach using historical market data, which in this case would be used to provide a "potential size" of the revenue but with a low confidence level; or it can be built using an actual pipeline of customers, and assumed success rate with those customers per year, which in this case will represent a more accurate view of the likely revenues because it is utilizing actual customer candidates and targets.

The values below are for demonstration purposes only and do not necessarily reflect an actual scenario:

Country Market Forecast (Data Service Provider)					
	Yr1	Yr2	Yr3	Yr4	Yr5
Eligible Subs with Smartphones	132,000,000	145,200,000	159,720,000	175,692,000	193,262,000
Assumed Annual Average Usage Per Sub (Units)	250	400	650	800	1,000
Price Per Unit ($)	0.025	0.023	0.021	0.019	0.017
Penetration of Eligible Subs	0.125%	2%	4%	5%	7%
Avg Subs Size in a Year	165,000	2,904,000	6,389,000	8,785,000	13,529,000
Forecasted Annual Revenue (Rounded)	$ 1,100,000	$ 26,800,000	$ 87,300,000	$ 133,600,000	$ 230,000,000

Table 3 – Example of Annual Revenue Projections for a Data Services Company

Table 4 – Profit and Loss. Once the overall cost and revenue projections are developed, you may create a summary profit and loss (P&L) view to reflect the total cost, total revenues, and forecasted earnings before tax, interest, depreciation, and amortization (EBITDA):

	Yr1	Yr2	Yr3	Yr4	Yr5
Total Annual Revenue	$1,100,000	$26,800,000	$87,300,000	$133,600,000	$230,000,000
Reward Cost (10% of Sales)	$110,000	$2,680,000	$8,730,000	$13,360,000	$23,000,000
Cost of Goods Sold - Data Access Cost (40% of Sales)	$440,000	$10,720,000	$34,920,000	$53,440,000	$92,000,000
Gross Profit	$550,000	$13,400,000	$43,650,000	$66,800,000	$115,000,000
External Cost (20% of Gross Margin)	$110,000	$2,680,000	$8,730,000	$13,360,000	$23,000,000
General Operational Cost (Expense/Capital)	$1,353,000	$4,958,000	$7,770,000	$15,540,000	$31,080,000
EBITDA	-$913,000	$5,762,000	$27,150,000	$37,900,000	$60,920,000

Table 4 – Example of a Summary P&L

Tracking Progress and Setting Revenue Goals

Once the revenue projections are developed for the first 12 months, three years, and beyond, the business can then, set revenue goals and measure the progress against them, month over month, quarterly, and annually. Some businesses may be more sensitive to seasonality, where a specific time of the year such as holidays, the beginning of school year, sporting events, weather changes, and other factors can generate a peak or a dip that is different to the rest of the year. Those facts need to be understood and considered when developing and tracking the financial performance.

Tracking revenue and financial goals is important, not only to monitor the progress of the business over time, but also to understand how and when certain business metrics will be met. These metrics can have a great impact on funding decisions, hiring of resources, marketing budgets, and other business operations. It can also tell you if you are growing as expected, or if you are facing challenges and need to adjust your business plan.

Payback Period

One of the key metrics that a business leader, investor, or a partner is interested in, is the time required for the business to recover the initial costs laid out in the venture. Some businesses may require three to five years to recoup the initial investment from the generated profit, while others may require only one to two years, and some even less.

However, certain investors are much less interested in this payback period metric as they see the value being developed in the company over time as much more significant than short-term profits. Their aims are much longer-term and the product or service is possibly intended to support another core product in the company, which can then have an associated measurable value. There are some technology startups that never generate profit yet produce significant payback through an exit event, typically in the form of an acquisition by another major player that sees important value in them for their organization. For example, I've been using Google Maps for many years with minimal to no intrusive ads appearing on my application; this state-of-the-art service is provided to me, and millions of others, for free. It has become part of our daily lives and a tool that we cannot do without. Such an approach is considered strategic and it ensures my continuous engagement with the Google suite of offerings and products.

The more these services are offered, the more likely I will continue to be attached to Google's services. Google Maps actually evolved from the acquisition of a very small startup, Where 2 Technologies, which was in its early development for map application services when they met with Google's founders and pitched their idea and its relevancy to Google's future offerings. While their technology had not made any money at the time, the fortune they earned through the Google integration and acquisition was significant. Other map services companies later got acquired as well to complement Google's internal development, such as the ZipDash that offered real-time traffic information. Many of those companies did not make a profit through selling their services directly to customers; rather, it was through the acquisition by Google. Google may not make any money directly through sales of the maps service application either, but more than a billion people a month will continue to use it alongside other Google services that do generate handsome amounts of revenue, such as advertising income, and thus it has indirect value. It also offers strategic benefits, keeping us on the Google platform for longer and preventing competitors from capturing market share. For example, when looking for a hotel or a restaurant, Google presents the results laid out on a map, which encourages us to use this search engine for convenience and also funnels us towards paid advertisers.

The key here is to determine your long-term strategic intent and comfort level with the payback period. Is your product intended to generate direct and quick profit for the company, or does it aim to provide an indirect value by providing support to other companies (potential buyers), or other internal offerings? If you are accepting capital from investors, expect to address this question and be able to set their expectations from the beginning on the likely horizon of the payback period and the form of exit that will realize returns on their investments.

Reaching the Breakeven Point

The breakeven point is the moment when the company can start covering its own operational costs without external borrowing or investments i.e., by sales revenue alone. Understanding the likely breakeven point helps you to plan and allocate sufficient budget or funds from third parties to support operations until the business becomes sustainable.

Some companies cover these initial costs from internal sources, especially if the business includes multiple offerings that are already generating income. The new product or service is considered an investment made by the company, and revenue generated from other offerings cover the launch. This is a typical practice with existing and established businesses.

Others, however, will need to self-fund, borrow, or sell shares of the company and invite investors to provide capital until the business is capable of standing on its own feet and reaching the breakeven point.

It is important to point out that some businesses may still decide to find external funding for other reasons, even when they are profitable. The leadership may decide to scale the operations at a faster rate than the current income allows, invest in new products, or grow into new territories where the current business revenue cannot support this.

Forecast your breakeven point and be sure you have a plan to fund the operations until the company can generate sufficient revenue to cover its own expenses. The forecast should include a risk margin so that, if the breakeven point is delayed, there will still be sufficient funds to carry the company forward.

Return on Investment (ROI) or Rate of Return (ROR): Measuring Value

Businesses not only want to realize financial returns in terms of recuring income and operational profit, but also from accumulating value in the business itself. This value increase in the business is developed through growth in current and projected revenue, in the value of its assets, in its intellectual properties, and in the value of its brand name. A good analogy for this would be a real estate property that may increase in value over time, where the sale price is two to three times its original purchased price, or of the original cost of development. Its increase in value may be calculated based on the rental income it generates, or on the overall market value increase of the real estate itself, or both. Similarly, a business that required $100,000 to launch could grow in value due to the increase in it is annual sales and annual net profit to many times its original launch cost.

There are other important metrics beyond revenue, profit, and speed of growth that can also increase a company's value. Increase in brand awareness, traffic, engagement, registration, revisits per customer, repeated sales, and more are all measures that can enhance it. As I've mentioned previously, WhatsApp was sold for more than $19 billion to Facebook, and at that time its annual revenue (estimated at $16 million according to Investopedia) was insignificant relative to its purchase price. The real value in the company was the number of "active users" and the growth in this metric, month over month. WhatsApp had close to 500 million subscribers, with significant daily traffic per user. Those numbers were also trending upward at a significantly high rate. Such metrics were of great interest and strategic value to Facebook.

While business leaders and investors always aim to increase the return on their investment, some can experience a decline in value. This can be associated with a consistent decline in sales and associated net profit, or the projection of a trending decline in revenue. This often occurs due to market changes, such as growing competition, shifts in customer behaviors, development of substitute products or services, and the inability of the business to evolve and continue to be relevant to its customers. These dips in value can occur during specific phases of business development, but they cannot be tolerated for long periods of time, as they could easily lead to a complete eradication of the business's value and an inability to operate. The dot-com market crash in 2001 was a great example of unsupported valuation. When the overvalued tech stock market peaked and the Nasdaq reached the 5,000-point level, investors saw this as a turning point. The overall market value started to head south without much support. Companies that were valued at billions of dollars without proven and reported revenues and projected growth rates started to drop at a steep rate, many by 90% or more. Others were sold for pennies a share, or simply closed their doors and shut down. iVillage, Looksmart, Infoseek, and many more lost the majority of their value as they could not prove their projected growth and ROI. While the tech market experienced steep losses in its value, those that were able to prove the soundness of their business model and demonstrate real growth survived. They later benefited from the reduction in competition from firms which failed as they didn't have sustainable business principles. To name a few of the winners, Amazon dropped from a $113 peak to a near $5.5 low, and later recovered to reach more than $3,200 per share in December 2020. Apple dropped from what is equivalent to a $1.35 peak, when considering stocks splits, to near $0.25 and recovered to reach more than $122 per share today. American Tower dropped from a $54 peak to a $0.70 low, and is now trading at more than $230 per share. Adobe Systems peaked at near $40 to drop close to $8 and today they trade at more than $500 a share. Intuit dropped from $45 to close to $11, reaching $355 a share today. All prices are adjusted to market splits.

You may be interested in some fun facts. An investment of $10,000 at the near lows in any of those companies during the dot-com crash would have yielded substantial wealth to those who owned it. As of January 2021, those shares would be worth as follow:

- Amazon - $5.5 million
- Apple - $4.9 million
- American Tower - $3.2 million
- Adobe Systems - $625,000
- Intuit - $300,000.

All this goes to show that companies which follow sound business principles can navigate difficult challenges, and can hold and generate value for their stakeholders, employees, and the societies in which they operate.

Return on Investment Calculations

The standard calculation for ROI is as follows:

(Current business value – initial investments) / initial investments

For example, if the initial value was $2 million, and the current value is $5 million, then

ROI = ($5M - $2M)/($2M) = 1.5 times or 150%.

The metrics for assessing business health do depend on the *type* of business. For example, wireless tower companies which build towers and poles, then lease out space on those structures to mobile companies for installing their wireless network equipment, use multiple measures to reflect the health of their business. In addition to total revenues and bottom line (net profits), the business value is also measured based on the size of its assets (number of towers, poles, rooftops, and other usable structures) and their occupancy rate. While the number of towers or owned properties is important, another critical metric is the average number of tenants (wireless carriers) that are leasing a spot on each tower, and the average number of installations per tower. The higher the average number is, the more confident the company is of securing recurring and efficient capital coming from long-term and committed customers. No one wants to see an empty tower with unleased spaces, since significant investment went into building and maintaining them. In this example, this measure indicates the efficiency of the capital spent in the company and the effectiveness of the operational teams, which are important to track in order to assess the long-term sustainability and profitability of the business.

There are also other metrics that need to be considered that can indicate new market conditions or recent changes in market environment, which may impact the value of the business in the future. These could be operational cost increases or decreases, pricing changes, creation of technologies that can increase or decrease the demand for your products, and the introduction of new government rules that could cause business restrictions or benefits. These all need to be evaluated when assessing the value of the business.

With that said, for most companies, sales and revenue growth, or lack thereof, continues to be considered the main dependable indicator when assessing the value of a business. When a change in direction is detected, whether it's positive or negative, it is important to understand the reasons behind the change and take action immediately to address risks or threats. Two businesses with the same income may have a completely different valuation due to the fact that one may be experiencing faster growth than the other, or in unfortunate circumstances where one is experiencing (or expecting to experience) a decline in its sales and revenue. When a company is experiencing growth, higher multiples of its sales or net profit can be used to estimate its value. When a company is experiencing a decline, or uncertainty, then its valuation should also reflect this risk

The Amazon Phenomenon – Profit Is Not the Only Measure of Value

In most businesses, running a positive net income is vital for its survival. There are a few exceptions, of course, where the company understands that necessary investments may produce losses until it reaches the mass market or desired milestones. For example, selling products at cost, or even below cost, can be a strategy used to build an initial customer base or develop market awareness of the business. However, this approach needs to have a calculated timeline and the funds available to support it. A business cannot indefinitely lose (in real money) and maintain its market presence. Amazon is one of these extreme examples. It was launched in 1994 and remained unprofitable for many years, bringing deep concerns for its investors and stakeholders. It finally turned a profit in late 2001, seven years later. The company survived by utilizing capital generated from a mix of resources, which included income from sales of products it carried, sales of company's stocks to the public, and external investors. Its ability to survive was not accidental, however, nor was the strategy foolish. Amazon had decided to focus on certain business elements, which they felt were critical to its operations and growth over short-term profit. Customer satisfaction, customer growth, market reach, and market awareness were all metrics that were prioritized over profit. The company was able to demonstrate to its leadership, investors, and stakeholders that its sales, customer base, brand awareness, and engagement were all growing, month over month and year over year, at a significant rate; it was enough to convince them that the company was growing in the desired direction and developing a competitive advantage that would see them taking control of the market.

As Amazon showed, when a business can demonstrate growth in customer size and revenue, profitability goals can be delayed until it picks up the necessary scale, momentum, and strength to navigate the competitive landscape and market forces. However, if a business fails to demonstrate revenue growth or successful market traction, then further investment may not be the answer to the problem, or at least it should consider adjusting its strategy before pouring more money into it. When facing challenges like this, many companies rightly pivot their operations by changing their products or services, adjusting their target segment of customers, and improving their sales and marketing plans. However, if the foundation of the business can no longer be supported, for example if the value offered is no longer of interest to the customer, such as Kodak films, Blockbuster video rentals, and RadioShack electronics supplies, and the business does not evolve to address these challenges, then the difficult decision to exit may be necessary.

Have a measure, or series of measures, in place to indicate the health of your business and attach a timeline goal to achieve it –whether it is sales size, revenues, new customer registrations, revisits by customers, utilization rate of your assets, or any other relevant metric to your business. Assess the metrics regularly and decide if the business is meeting your goals; this will help you to make further strategic decisions. Failing to set metrics to guide your progress is the equivalent of sailing without a destination.

LEADER'S SUCCESS QUESTIONS

Principle 8: Know Your Financials (Revenues)
What is Your Revenue Model and How Will You Track Your Progress?

Ask yourself the following questions in order to develop robust business revenue and value projections.

1. How many customers do you think will use the product per a given period, and how often?

2. What are the monthly revenue projections for the first one to three years? Some choose to project five years or more. When developing long-term projections, the business must acknowledge that there will be market fluctuations and unforeseen risks, and therefore projections beyond the first three years should be viewed with lower confidence levels.

3. Are the sales and revenue projections defendable? Why would anyone believe the projections? Can you explain your method and support your assumptions with credible data? Do an honest self-assessment of the assumptions and facts used, and be sure you convince yourself as a business leader before you try to convince others, such as investors or partners.

4. How profitable is the business and when do you reach the breakeven point? What is your desired and projected rate of return? What is your payback period as an associated strategy?

5. What measures do you need to have in place to validate meeting your desired projections? This might include sales, customers, return visits, and other measures that are relevant to your business's revenue.

Principle 9

Build Your Go-To-Market Strategy:

What Will Your Initial and Full Market Rollout Look Like?

Your go-to-market strategy (GTM) is an opportunity for the organization to shine, compete, and capture market share. It is one of the most fundamental parts of your business plan and launch strategy. It describes how you are going to reach your customers, where you will sell, how fast you will grow, and whether partners are part of the plan. It also describes how to address the competition, what the short-term initial product or service will look like, the evolution of the long-term roadmap, and what price point to rollout with. The revenue projections and pricing determination, as we discussed in the earlier sections, all feed into your GTM plan. The GTM may also include details of whether customer trials are needed as part of the purchasing decision, what markets you will engage first, and how you are going to reach those markets, such as through an online presence, a physical presence, channel partners, or a hybrid model that includes all three.

All these elements are fundamental in determining your customer reach strategy. While every business in every industry is different, focusing on and understanding those points can make the difference between successfully meeting your target goals, or missing them. They should all be examined carefully, in turn, and you and your supporting team should make strategic decisions about how to tackle each one.

The main elements of a go-to-market strategy are:

1. Determining your target markets. These are the markets you want to sell to, and the segments you want to focus on, as discussed in Principle 2 (Understand Your Customer).

Your GTM plan should include the countries you will operate within and, for each country, which city, town, or neighborhood you will market your offering to, either through a physical or online presence. Each market may have its own unique characteristics that are affected by culture, customers' purchasing power, interest level in your offering, environment and seasonality factors, and government regulations. Those variations or factors should be recognized when developing the GTM. In addition, you should define your customer segments in each market. For consumer segments, a combination of age, gender, education, ethnicity, language, income level, and other characteristics are relevant. For business-to-business customers, you may want to describe industry categories such as retail, healthcare, auto, insurance, technology, consumer devices, travel, education, consulting etc. Therefore, determining where you intend to sell your products and services, and to which segments you will be selling, should be detailed and defined in your initial GTM plan.

2. The rollout speed and sequence of regions. This means planning the rollout timeline within each selected market and for each segment.

In addition to determining the regions you are going to operate in and the segments you will serve, your plan should consider a sequencing order for each of those regions and segments and the speed of rollout. You may not be able, even if you wanted, to target all markets at once. There are practical limitations and barriers that can reduce your speed and reach. The cost of rollout in multiple markets or segments at once may be prohibitive, and the uniqueness of each region and segment may also require different strategies. Furthermore, a plan that includes sequencing of markets would offer a useful learning curve that can support optimizing the overall GTM as you grow and expand. Your plan should include the prioritization and timeline for each of the regions and the segments you will start with and those you are growing toward.

3. Customer acquisition process. This defines how your customer accesses your product or service (physical location, online/mobile, both).

Your plan needs to define how customers will access your products. Is it through direct visit to store, submitting an application, or online? Do they need to register, get prequalified, or sign up? What payment methods do you accept and can they pay online, offline, or both? Will they receive the product immediately, or will it be delivered or emailed depending on what your offer is? The process of buying and receiving the product must get defined in your GTM as it will also influence your organizational structure, including the design of your stores if you need a physical location, and the design of your digital platforms.

4. Channel partners. Will you need a channel partner to help you sell the product in a given region or online?

If a channel partner is needed, then you should determine what the required characteristics of this partner should be, such as their strengths in the industry of your focus, their customer reach, market recognition, brand trust, a strong online presence with a digital platform, or the ability to process customer payments etc. You will need to define the commercial terms under which you will work with the partner, and the cost of the partnership. Some partners may require special digital integration to work with you; others may require dedicated resources to manage the relationship. In addition, your partnership terms will need to discuss the revenue share, or the fixed price for their services, as they will have their own costs and profit requirements. Some channel partners will incur less costs than others if you have the same target customers. In this instance, they can achieve economies of scale in working with you as they can use their existing sales and marketing structure to offer your products for only a minimal increase in their sales cost. Other areas of collaboration can be around services and support of the customers. You will need to decide, based on their capabilities, if you want them to handle the support, repairs, and returns, and answer customer complaints, or if you would handle these. You'll also need to decide what communication model is used if they do handle customer support. There are other terms that also gets included in a partnership agreement, including who should collect customer payments, and when revenue should be shared in a given month. There should also be agreement on a framework to address disputes related to sales, cost, and revenue between the partners. A legally-approved agreement is recommended, where each party receives independent legal advice and representation when structuring and reviewing it before signature. While partners will have a cost, they can offer great advantages such as faster reach to new markets, less cost to establish a new presence, credibility in a given market which you may be able to leverage, and access to existing sales infrastructure, such as stores or digital platforms that can support your products.

5. Building your marketing strategy. In your marketing plan, you'll need to allocate a marketing budget, develop marketing messaging, decide what media to use, and what digital and non-digital forms you will employ.

Your plan may include a combination of digital online channels, mobile, email, text, newspapers, local marketers, broadcasters, conferences, and other relevant channels. For each, you should determine which media you want to use, and the size of the budget to allocate. Identify the most effective channels, media, and marketing tools for your business and how they will bring awareness to your offering and its differentiation. Some parts of your strategic plan and your marketing spend will likely prove to be more effective to your business than others. There is a famous saying in marketing, "I know that only half of my spending is working. I just don't know which half it is." While you only want to spend resources on the marketing media that is most effective to your business, it will likely require time and money to identify the best and most appropriate channels before expanding and optimizing your marketing spending. Seek advice from internal and external resources available to you, and experiment with a small budget first before committing to a given media or marketing channel.

6. Defining your sales strategy. This is all about building your sales team and their structure to optimize your approach.

Your sales strategy will include the type of sales skills your team need to have based on your industry and products, how many people you need, and how they are going to be structured. It also includes the process by which you acquire customers and sell your products or services. For example, your sales plan may require having one-to-one sales personnel to pitch the product or the service, answer questions, and provide guidance on buying choices. For example, a real estate business will still need employees with specialist knowledge to meet with clients in person and "close" the purchase. You may also need dedicated sales representatives for business-to-business offerings, such as when selling technical solutions. Large tech companies such as Cisco, IBM, and Ericsson, and many smaller tech vendors, have sales teams supported by multiple marketing channels to generate sales with target customers. Other sales models may be fully automated, and you can direct the customer to buy the product or service through a fully digital flow, such as when buying Microsoft Windows products online, doing your taxes using TurboTax, or buying a book from Amazon using their digital ecommerce platforms. Some sales models require a hybrid model, where a salesperson is needed to address customer questions or to initiate the process of acquiring new customers. Once the customer agrees to purchase, they can then be directed to the digital platform to complete the sale. The sales strategy will also include decisions made on the pricing structure and price points of the product. It may include selling a basic version of the product or service at a very low price, or for free, to gather as many potential customers as possible, and then offering an upgrade or a premium version for a fee. The premium version of the product could be purchased digitally or by contacting sales representatives. This sales model is often used with online gaming products where a basic subscription is offered for free, but then players are encouraged to buy additional digital products or essential tools to compete in the game for a fee, using an online payment form. The sales structure is changing in almost every industry due to the technological advancements that have become available over digital platforms and payment processing infrastructure. There are now fully-automated digital platforms that allow sales of items that were not possible before. Selling vehicles, for example, has radically changed over the years. There are now companies offering an almost fully-automated sales process, from the time you review the car's interior, exterior, condition, and its history, to selecting to buy, to financing it, and then picking it up from a nearby location or having it delivered to your home. Carvana, Vroom, and CarMax are few examples of this significant shift in the sales model to digital platforms in the automotive industry, where it used to be heavily dependent on one-to-one sales staff. With that said, human connection is still considered vital in sales, especially in large business-to-business deals, where a skilled and competent salesperson is required to develop trust and establish a relationship with customers to manage their evolving needs.

The sales process, whether it is digitally enabled, utilizes a hands-on salesperson, or is a hybrid of the two, typically goes through a cycle. Depending on the industry and the type of product, this cycle can be short, with a decision made in a few minutes, to a more complex sales cycle where the process can take several months. For example, the typical business-to-business sales cycle starts with establishing prospects which help with the development of what is called a "pipeline." The prospects are identified through market research using digital search and digital marketing, existing connections, sales events, advertisements on multiple relevant channels, and industry conferences. The goal is to identify a large number of customers that are likely to buy as they fit the criteria defined for the target segment of your business. Once identified, the process of researching their needs and pain points starts in order to reach out and develop awareness of what value you can add to their business. The goal is to generate interest by ensuring customers want to learn more and start developing an appreciation for your product and how it addresses a value they care about. Once interest is developed, the next step is to convert this to a purchasing decision by developing relationships and common ground with the customer's stakeholders, influencers, and decision makers in the company. This engagement, if successful, concludes with an agreement to purchase.

This is really not the end of the process but rather the true beginning of a new cycle that strengthens the relationship and supports engagement in post-sales activities, such as delivering on what was promised and servicing the customer account to their satisfaction. This is important, not only to the success of the existing deal, but also to ensure you are developing more opportunities in the future with customers. Throughout the sales cycle, establishing trust and demonstrating your competitive advantage, supported by a well-researched pricing structure, can help in shortening the sales cycle and assisting the customer in making their decision faster.

Buying decisions in business-to-business sales are influenced by many factors but, in my experience, it is usually dependent on three things: how closely the product satisfies the business's needs, the level of trust developed in the relationship with the company and its dependability, and pricing. The product must address an important problem that will either save them money or increase revenues. Customers that become convinced of the product's value, whether it increases their efficiencies, effectiveness, or competitiveness in gaining market share, will have a high level of interest. Trust between the two parties, on the other hand, can play a great role in shortening the sales cycle and the response time by the customer. Trusting the quality of the product, and the reliability of the company in its ability to deliver the service or product, helps strengthen the position of the sales team in closing a deal. In addition, the pricing and pricing structure is a powerful sales tool and it should take into account the ability of the customer to pay and the dedicated budget for the solution. In certain cases, a request for proposal (RFP) may be required, where multiple companies can respond to compete with your product. At the end of the day, people make decisions that leave them feeling satisfied that they have made the right call for their organization. Generating this feeling in a customer is dependent on having a skilled sales team to deliver the right message and/or a well-designed sales digital platform supported with innovative marketing messaging.

The consumer sales process goes through similar steps, but the time taken, and the tools used to move from one step to another, are different. Consumer sales are becoming more dependent on customers making their own decisions and leveraging the available tools provided to them by the company (i.e., their website), and less on actual live sales personnel. Smart digital sales tools are making significant technological advancements using machine learning and artificial intelligence (AI) that can help identify prospects and offer them the product or service at the right time and via media such as mobile, streaming TV networks, and web applications. Take advantage of technology, but be mindful of the type of business you are in and what sales' strategy and sales' skills are required for your success.

7. Preparing your sales' contracts and legal agreements. This is the essential step of developing the documents and terms and conditions for customer sales.

Many of today's products and services are governed and regulated by certain government rules and policies. Many of those policies are meant to protect the privacy, safety, security, and the rights of customers. Whether you sell a consumer product over a digital platform, or in a store, or whether you are offering a business-to-business solution, a legal sales agreement that includes terms and conditions is likely needed. You should engage a legal entity to help you construct this agreement. However, the initial work that goes into sketching out the agreement should start with you. You will be expected to draft a description of the product or service you are delivering, what your responsibilities are, what the customers' responsibilities are, your commitment to customer support during the product delivery and post-sales (as applicable), the price and pricing structure, and payment terms details. Once you have a general draft, you can engage a legal representative to produce a generic agreement for every sale. Some types of sales may require some customization, and business lawyers are able to produce these agreements with the understanding that some parts can be adjusted for this purpose. Prepare this agreement as close as possible to the time you need to engage in the market so that it's up to date, but don't leave it to the last minute as it can prolong your sales cycle unnecessarily.

8. Defining your post-sales support model. How will you service your customers? Would it be over a phone call, via store visits, or remotely, and what would the response time be to resolve their complaints, fix problems, and satisfy their needs?

The needs of every customer are different. However, whether it is a consumer product, or a business product, customers expect that you will be there after they buy the product in order to address possible issues, returns, fixes, and other potential maintenance work that may be needed. Customers need to have confidence that you will still be there next year, and even for the next five years, when making their buying decision. Some products, however, require urgent and timely support and therefore the customer will be demanding something called a service level agreement (SLA), which will describe the process and the expected time needed to resolve problems raised, based on the severity of the issue. For example, imagine telecommunication equipment went down during a busy hour in a major city where hundreds of thousands of people and businesses were impacted. This problem cannot be addressed in a 24-hour response time window. The telecoms provider would expect it to be addressed immediately at the highest levels and given urgent attention to be tackled within minutes. Every second that passes will prevent communication between critical operations such as 911, which could lead to a life and death moment. Furthermore, the revenue losses from every minute that the network is down would be huge and could have a long-term financial impact as it could damage the telecoms company's brand and customer confidence. When the network is down, businesses stop conducting transactions over the network and that causes substantial monetary losses. Therefore, for products that require a critical and urgent response time, you will need to be prepared and have the organizational structure to offer this level of support. Not all products required such an urgent response, however. For example, if my central heating breaks down, I may be able to wait for a day or two until it is fixed, even though it is causing inconvenience. I can possibly wear a few extra layers or purchase a portable space heater if the weather was cold. Obviously, having your heating break down while living in Alberta, Canada, in the middle of a bitterly cold winter is significantly more urgent than if you lived in San Diego, CA. Therefore, expect your customers to ask for a service level and response time that corresponds to the criticality of the product and service you are selling.

9. Cloud-hosting decisions for digital products and services. Should you host your application in the public shared cloud, on your own organization's dedicated servers, or within customer's premises using their own servers?

If you are selling a digital product or service, one of the decisions you will need to make is whether you install the product in the cloud and provide secure access to customers, install it on your own servers within your own company, or install it on the customer's own computer. As part of your GTM strategy, you will need to make that decision based on the nature of your product, the customer's requirements, cost, scalability, reliability, availability, and your ability to perform maintenance and support post sale. In addition, you will need to consider privacy rules and each country's regulations related to hosting, as some countries allow local data hosting to keep their data from leaving their borders. Many financial institutions, however, still require that applications are hosted within their own hosting centers, and they are willing to spend the capital on establishing the computing infrastructure and maintaining it to keep control over performance and security. After all, their infrastructure holds records for hundreds of billions of dollars and they need to ensure they have full control of the data. With that said, major cloud-hosting companies such as AWS, Azure, Google Cloud, IBM, Oracle, and Salesforce, among many others, have developed and employed the highest levels of security, reliability, and scalability systems at a reasonable cost. Therefore, we are seeing even highly-sensitive businesses, including those in the financial industry, utilizing these reliable cloud-hosting companies for their data and applications. You may also need to consider the location of where you would host your application relative to the country you are serving. The delays are typically very small or unnoticeable for most of the applications; however, it is good practice to consider hosting your application within close proximity to your customers. You may need to run tests to measure the experience and determine if you should host from a single location regardless of where your customers are, or host in multiple global regions, if needed. When you start a business, it is almost always good practice to start with a single cloud company that is reliable and then decide to grow or change based on your evolving needs.

10. Creating the financial business case. This includes pricing structure and projected revenue targets from each region you are selling within, coupled with your projected associated cost.

By now you will have developed your pricing model, revenue projections, and built a good understanding of your costs, including your initial launch costs, your ongoing variable and fixed costs, and any additional costs associated with each sale, if any. You should include a version of your financial business case as part of your GTM strategy that will reflect the health of your company's likely profit and loss (P&L) statement. The business case can be developed using a simple spreadsheet such as Microsoft Excel to list your main financial elements, including planned pricing, sales, revenue, costs, and projected net income.

The sheet can be structured into sections as follows:

1. In the first section, you may include an itemized projection of your sales units or transactions, month over month, or as frequently as applicable to your business. This can also represent the number of customers, unique registrations, unique visits, or repeated purchases. This will enable you to pin down the likely number of transactions that will generate revenue.

2. Then, in a separate itemized section below it, you may add your associated revenue based on the projected sales, in addition to any other revenues that are projected to be earned from other possible sources, such as earned interest, maintenances, real estate income, subscriptions, or equipment rental. Some businesses receive the sales income immediately, while others may have a 30- or 60-day deferred period. Others may be dependent on other conditions, such as payment upon delivery, upon installation, or other business conditions that need to be satisfied in order to earn the revenue. You may want to add these delays in revenue to the spreadsheet if you are interested in the projection of the cash view in order to determine the capital required to operate, month over month.

3. The third section should include your associated projected cost per each period or per each sale. This should include your variable and fixed costs, as discussed earlier, and any bank payment or loan obligations that you will need to pay out.

4. The fourth section reflects the projected profit or losses that you would realize per each period. This is simple math of subtracting the revenues from the cost per period.

5. A fifth section may include the effects of a risk factor such as 10 to 30%, depending on your confidence level in the estimated costs and revenues. The risk factor would reduce the overall bottom line by the reserved risk.

Let's look at the initial rollout and full rollout stages in more detail.

Initial rollout

Companies pioneering innovative products or services usually include a small, friendly rollout (or product trial) with an initial set of customers, giving them the opportunity to gain significant insight on the product and customer responses to the offering. Offer incentives if they need it in order to join, such as a price discount, limited exclusivity in a market, or the ability to influence the product's functionality to their advantage (as long as the benefit can be extended to others). Such validation should be part of every GTM launching new products, as it's an invaluable way to learn more about your product's value and how customers will engage, as well as usability, functionality, and how closely it aligns with the target market's identified needs. You can also gain insight on customer response to the pricing structure and their willingness to pay for it. Research through focus groups can also provide valuable insight before a full rollout. However, a focus group "customer" who would theoretically buy your product at a particular price point will always be a more unreliable measure of success than an actual customer putting in an order.

If a small scale, friendly trial is not possible due to the nature of the product, then start with a throttled rollout or a slow one, where you may still gain the needed feedback to optimize your GTM. The slow rollout enables you to optimize the product or fix any issues you discover that weren't apparent during initial testing. Whether it is safety issues, quality problems, or a price point that seems to be unfriendly to the desired audience, insight at this early stage is extremely valuable to the success of the business.

Plan your initial rollout and define it over a timeline that can be measured in days, weeks, or months. Once you have absorbed the feedback and finessed your offering, you can launch on solid ground before scaling up your market reach.

Full Market Rollout

Once the initial rollout or trial is concluded and necessary adjustments are made to your strategy, the product is ready for a full-scale launch. The market rollout plans should define the pace and speed of the launch; what cities, states, and countries to launch in, with a timeline for each, and whether they will be geographically phased; and, finally, whether each will include a physical presence where an actual store, show room, or customer visiting center will be needed, or only an online presence such as a website or mobile application. Each of those decisions require a defined organizational structure and resources and capital to execute.

Obviously, launching within a single city or country is less challenging than launching internationally, where country-specific taxes and special licensing can require more planning and may complicate the operations and associated cost of rollout. For certain offerings, such as digital products, the availability of cloud hosting and global digital connectivity have reduced the complexity of scaling and delivering offerings globally. Many governments throughout the world today realize the value of attracting external investments and have introduced regulations that make it easier for overseas businesses to operate within their region or from another international destination. Nevertheless, if you have plans to roll out your offering internationally, be sure to understand the various regulations that may impact your ability to sell in various destinations. Some countries may require a local presence, or that you register to incorporate; others may require you have a local partner or agent. There may also be various hidden costs, such as customer taxes, sales taxes, and income taxes in the particular country you are launching in, that need to be accounted for. Money flow in and out of a given country and transfers can also have restrictions, depending on where you operate. Some countries may only allow transfers of money based on a trail of supporting documents that prove that local taxes have been paid for each transaction.

As the team execute the full market rollout plan, key measurement indicators should be put in place ahead of time, which help to gauge success in order to guide further optimization of decisions. Customer acquisition volume, sales, revenues, registrations, etc. are common measures that can provide the necessary feedback about your business.

Learning from Existing Models

When developing your GTM, try to learn from existing successful models and the experiences of others in the business you are launching in. You are not looking to replicate what they have done, rather learn from them and add your own unique model or processes to further differentiate yourself.

If you are developing a high-tech business, the startup community is rich with resources and there are many experienced advisors that can lend a hand to support your efforts. Incubators, accelerators, and other corporate-driven events are available to help. Also, consider inviting advisors with relevant experience, skills, and knowledge in the industry to join you, as they can offer insights and recommendations to enhance your strategic plan. Bounce your ideas around with trusted individuals who are willing to review and help your business GTM. There is tremendous value to be gained from this sort of feedback.

Your go-to-market strategy is the way in which you can significantly differentiate your business and gain an edge in acquiring market share. Plan and execute carefully to reap its rewards.

LEADER'S SUCCESS QUESTIONS

Principle 9: Build Your Go-To-Market Strategy
What Will Your Initial and Full Market Rollout Look Like?

Ask yourself the following questions to develop a robust go-to-market strategy:

1. What is your initial rollout plan? What is the expansion plan or full market rollout plan? Define the initial rollout, including trials to engage friendly users or a small set of customers. Offer incentives if they need it.

2. What is your full business case? Leveraging the pricing, cost, and revenue projections you have decided on earlier, develop a version of your business case to help you understand the projected financial health of the organization per a given region and given period.

3. What are the price points that you will initially rollout with, and when will you consider changing them? How are you going to measure customer response to your price points?

4. What are your target markets? Where will you offer your product to customers and in what markets and segments?

5. How fast will you expand to include additional markets and customer segments? During the full market rollout, is a trial needed in every new market, or would the results

from previous trials be sufficient to prove the value to customers in each new region?

6. What is your customer acquisition process? How are you going to sell your product when ready? Define how customers are going to acquire and receive it i.e., what is the customer buying process? Would the business need a physical presence, online presence, partnership channels, or a hybrid model?

7. Do you need a channel partner? If so, what are the criteria that will help you in selecting one, and in what markets? Also, what terms and conditions, including revenue sharing, are you willing to offer to the partner, and who will service and support the acquired customers?

8. What is your marketing strategy? Where are the best places to market the business? How will people find out about your product? What marketing channels are you going to use?

9. What is your sales strategy? Who will sell your products and what is the structure of this team? What is your sales process and how long is your expected sales cycle? Do you have legally reviewed and approved customer agreements with terms and conditions defined and ready to use when closing a deal?

10. What is your post-sales support and maintenance plan, and what are your planned customer service operations? How will you address urgent and critical product issues vs. non urgent raised issues? For business-to-business sales, your commitment will need to be described in a service level agreement (SLA), which should be reviewed by a professional legal expert, and which should reflect your terms for response time to the delivery, installation, and for solving product issues.

11. For digital products, what is your cloud-hosting strategy? Will you be hosting your application at a highly-reliable commercial cloud company, within your own servers, or installed within each customer's servers?

12. How do you measure success? Define your key success indicators, including customer satisfaction, sales, customer acquisition, customer engagement, repeat sales, and growth milestones, to keep aware of what works best and what needs to change. Understanding the performance of your GTM helps you optimize and adjust it as you go along; for example, your spending on various media channels.

Principle 10

Secure Your Funding:

What Capital Do You Need and How Will You Find Investors?

For some business leaders or business owners, their funding source is internal - either self-funded from their own capital or ideally from their own sales to customers. However, for many, the capital they need to launch their business in a timely manner and capture the market opportunity may necessitate external funding sources.

Let me start by saying that funding your business through external money is going to be hard, and may prove to be one of the most challenging steps in launching your company. Asking a person or an entity to share the risk with you by providing capital, while having very little influence on operational decisions, is not a simple ask. Statistically, according to various online venture tracking sources, less than 1% of startups manage to receive some external funding and only about 0.05% actually raise serious amounts through venture capital. This is a very difficult task and you will need to be prepared for it mentally, physically, and organizationally. The discussion below can help you to improve your chances of securing funding to launch or grow your venture.

Internal Funding

Internal funding, whether from the individual or the business, is typically preferred as the business will continue to fully control decisions and direction with less influence from external investors. If a new product was proposed within a large organization, this would typically kickstart a process where it would go through several internal gate reviews before securing the funds required to launch. Usually, the business leader, or the product manager, would develop a business plan for the new product or service and present it to a special committee. This would help it to pass an initial assessment review, after which there would be several more committees to ensure the plan has been thought through and has considered the various market risk factors involved before approval. Gate reviews from business leadership, product, engineering, strategy, finance, marketing, sales, and legal teams are expected to be passed in order to gather approvals. Those reviews usually offer great benefits and insights as the product leader has to respond to difficult questions. This process can result in approval of all the funds requested, partial funding with timelines based on meeting milestones, a request to optimize the plan in order to address valid risk points, or they may decline to fund the product and cancel the proposed operation.

Small businesses, on the other hand, are mostly self-funded by a founder or a business owner. They may not have the benefit of expert feedback that a large corporate reviewing process provides. Therefore, identifying trusted and competent individuals from a friendly network can be helpful and is encouraged. This may include inviting advisors, friends, and family members to review the business plan and offer their input before you commit to funding it.

External Funding

Some businesses may require a larger amount of money to launch than what is available from the business owners or the organization. Therefore, external sources for funding might be required. Some companies may also make the strategic choice to borrow all or part of the funds, even if they have sufficient capital for the business. This might be because they want to keep more cash reserves, grow their operations at a faster rate, or invite strategic external investors whose participation will increase the chance of success due to their contacts, skills, knowledge, or even their potential for being a customer in addition to an investor.

There are multiple parts to funding through external sources. First, you need to understand that the maturity of your product or business as a whole can make a significant difference in your ability to secure investment money. If you are at the idea level, there will be a much smaller pool of interested investors than if you already have a prototype or an MVP developed. Moreover, interest will increase with the level of demonstrable market traction, such as sales, customer sign up, or signed contracts. The more mature your product is, and the more market interest you can show through sales and revenue, the more quality investors you will attract and the larger the fund sizes you will be able to secure. For example, at idea or concept level you may attract individual investors and a few institutions that are willing to invest in multiples of tens of thousands to low hundreds of thousands of dollars. With a mature prototype or MVP with demonstrable customer traction or sales, you can attract hundreds of thousands to millions of dollars, typically up to $5 - 8 million. There are always exceptions and we may hear stories of a business that has secured millions of dollars at the concept phase. I can assure you, however, those are few and far between, and you should not set your expectations based on these outliers. There is likely more to those stories, which leads me to the second major factor that can affect your ability to raise funds: the team. Investors often need to be convinced by, and trust, you and the team as much as the idea, product, or business. The more competent the team is and the more convincing based on their skills, experience, and education, the higher your chances are of striking a deal. It helps if this can be testified to by trusted and high-profile figures in the industry or powerful customers. This is true of many of the most success-ful companies that managed to secure $500,000 or more in their early stages, such as Facebook and LinkedIn. Third is the quality of your pitch. The better your ability to communicate the value of your business in addressing a major pain point, the higher your chance of success. But if you do decide to go down this route, then get ready for a long process. It will take a significant number of pitches, meetings, and networking events over an extended num-ber of weeks and probably months to secure your first meaningful investment.

Let's take a closer look at possible external funding sources, which can vary from being borrowed from a financial institution such as a bank, to being funded through friends and family, angel investing groups, crowdfunding, or venture capital (VC).

Each has its own advantages and conditions to meet in order to be considered. Not all investors will be willing to lend, as their policies and rules may dictate the phase and terms that would allow them to get involved with the business.

Direct Lending from Banks

Depending on the business, a bank loan may be an appropriate option. However, loans from banks can be difficult for startups since, typically, there is no history to use to support meeting the conditions of the bank. Most banks require records of operations and performance for multiple years, including tax filings, which may not be available for a new business.

However, if a business has a guaranteed contract from major customers, the business owner may be able to obtain a direct loan guaranteed by the future revenue of the project, making it a low-risk venture. You may consider providing collateral for the loan if this is needed by the bank to reduce debtor's risk.

Friends and Family

Friends and family can be a great source of funds to launch a business. They have a real interest in your success, and those who can help may agree to provide the funds if they are able to financially and trust your business capabilities. However, you must keep in mind that while they might be willing to help you, their contribution must not make up a significant portion of their own wealth or assets. You never want to be responsible for someone losing their life's savings, as there is always this possibility. The support you receive from family or friends must not hurt or disrupt their life or future stability. This could become an additional burden on you on top of the responsibility of launching and running a business. The borrowed funds from family and friends may take several forms. It could simply be a loan with a low interest rate, or it could take the shape of an investment where they may expect to own a small stake in the business as a result of funding your venture. If you do well, they too will see a rise in value of their stake and get rewarded accordingly. You need to be clear on what basis you are borrowing funds from friends and family. For example, if you are giving them business shares, you and they should have a clear expectation of how much you are willing to part with, which can also be documented in a financial instrument such as a loan note.

Angel Investors

There are quite a few individuals who invest in promising businesses in order to be part of their future success. Many of them are experienced and ask good questions in the process of verifying the investment opportunity, providing much needed guidance and critique. An angel investor can be an individual or part of an angel network who meet regularly and review opportunities that are pitched to them. They typically provide tens to hundreds of thousands of dollars per deal. Similar to friends and family, avoid taking an investment which makes up the majority of their portfolio. The US government has regulated such investments and angel investors must be prequalified and meet certain conditions. Typically, the net worth of the individual is required to be north of $1 million (without counting their own dwelling), or the annual household income should be greater than $250,000. The typical angel investor is looking to make anywhere between five and 10 times their initial investment. Some end up making significantly more.

The investment can be made in several forms. For example, it might be a direct stock purchase with a known share price value that is mutually agreed upon based on the company's valuation of itself, or by a third party. For example, if the business believes it has a value of $1 million based in its current revenue, or the projected revenue using realistic assumptions, and if the company has 10 million shares issued, then a $100,000 investment would mean the angel would own 1,000,000 shares of the company at a price of $0.10 per share.

The second form of investment is as a convertible note, where the loan amount can be converted to shares once certain terms are met, such as after a certain period has passed on the note, or if the company receives a major large investment round from a financial institution such as a venture capital firm. Typically, the note would carry a reasonable assigned interest rate, anywhere from 5 to 10%, where the norm is somewhere in between, such as 7% or 8%, and a discount rate to the advantage of the investor when compared with the market value at the time of converting the note to shares. The discount rate is applied as an incentive and typically ranges between 20 to 30%. A convertible note is often used when the valuation cannot be agreed upon by both parties, or if there isn't sufficient data to confirm a reasonable valuation for the company. It then becomes a good financial instrument as the angel and the business leader can agree to defer the valuation until they have more information. As an example, an angel investor might provide $100,000 in the form of a convertible note that holds a 20% discount and 8% annual simple interest. Assume the company has 10 million shares issued. If, after one year, the company receives a major funding round from venture capital with a post money valuation of $5 million, then the note would convert at a discounted share price of $0.40 per share using its 20% discount, and the angel would own 270,000 shares or 2.7% of the company.

Should the company do well, the valuation of the company may increase many times over, and so would the value of the angel's investment. As a business owner, it is not a good idea to offer any personal collateral against such investments, as companies and products can fail through no fault of your own. Investors are willing to take on a certain amount of risk as they hope to receive a rate of return that is equal to several times the value of the initial investment when the business succeeds, but can also lose the investment entirely if the business does not. It is a risk that both parties must understand, but with the intention to make all possible efforts to succeed. Therefore, angel investors typically invest not only in the idea presented, but also in the leader who will spearhead the execution of the business.

Crowdfunding Using Online Digital Platforms

Crowdfunding digital platforms provide everyone with the opportunity to present their business opportunity to a large number of interested people and invite them to invest. This approach was not possible in the past due to restrictive regulations which did not allow small investors to participate in offering capital in exchange for equity. Investors needed to have a high level of income, greater than $250,000 in the US, or own significant assets greater than $1 million in order to engage in such transactions. But, with changes in government regulations allowing companies to sell equity for even a small amount of capital, and the development of digital platforms that allow small investors to connect with companies to raise funds, many business founders have started to utilize these new funding channels. Typically, the platform gets a small percentage of the raised funds and some have a small fee for listing the opportunity. In exchange, investors get an opportunity to own part of a promising company they funded based on an agreed number of shares. The business can raise funds up to a maximum as capped by the law in the particular country. The maximum in the US, for example, as of 2020 is $1.07 million and, in Canada, it's C$1.5 million. In many countries, including across Europe, Australia, Russia, Asia, and South America, the maximum amount can reach several millions of dollars, allowing the company to kick off its operations comfortably. You should search the rules within each country to understand the opportunities to use crowdfunding and the associated regulations. Crowdfunding is being utilized today not only for investment opportunities to raise funds in exchange for equity, but also to fund nonprofit projects, personal projects, and other various charitable causes. In this case, the funds are not offered in exchange for equity, but rather they can be a pure donation in exchange of preordering the offered product, or sometimes some personalized gifts. You will need to review the terms of the crowdfunding platform carefully and determine which may be suitable for your business before committing.

Incubators and Accelerators

For high-tech startups, there is a class of funding institutions that is focused on supporting and investing in the very early stages of a company. They are called incubators and, for a slightly more mature business, there are accelerators. Their name says it all. They tend to get involved at the idea stage and help entrepreneurs to meet early milestones that are needed to advance the idea from concept to business. They support strategic introductions to experts, advisors, and potential customers. They also help in pre-

paring the early-stage venture for the next funding round opportunity by presenting the company to serious angel investors or
venture capital institutions, including family offices and private
equity firms. They provide some seed money that ranges between
$20,000 and $150,000, with an average of $30,000 to $50,000.
They may also include services such as a workspace, internet, and
coupon certificates for free cloud hosting of your digital application or platform at some of the primary cloud-hosting companies
such as AWS, Azzure, IBM, or Google. An incubator or accelerator program lasts between two and six months and typically
includes solidifying the business pitch over multiple iterations and
presenting it at pitching events. It concludes with a graduating
pitch-day mega event where many strategic investors, potential
customers, and industry experts attend to watch the final pitch by
each of the participating companies. Many investors often extend
investment opportunities to buy equity in the presented companies during the final pitch day event. Others choose to engage
further after the pitch-day event with startups that piqued their
interest. If the startup is in an industry that meets their investment
criteria, they work to further evaluate the company and determine
if they should extend an investment offer. Be selective, and inves-
tigate the accelerator program before committing to one. Review
the previous events and the success rate of their previous portfolio
of hosted companies. This should give you an indication of the
value you may obtain from engaging with them. Some of the well-
known names are Y-Combinators, Plug and Play, Techstars, 500
Startups, and many more that can be found in a Google search.
Examples of companies that were assisted by accelerators are
Airbnb, Dropbox, Zipline, and Outreach.

Venture Capital

Typically, venture capital (VC) companies invest in businesses that have already been launched and have demonstrated market traction, either in terms of customer acquisition, customer engagement, or sales and revenue growth. VCs can provide a great boost to the business if they choose to invest. Their investment amount can vary from low hundreds of thousands to tens of millions of dollars or more. The investment round typically results in diluting the founder and other existing investors' shares between 25% and 35% in each round. VCs typically invest through multiple rounds, and will likely continue to invest again as long as the company demonstrates success and meets its set milestones and key measures.

As with almost all funding institutions, when investing they focus on fundamental elements that help them make a robust decision. When meeting with a candidate company, the VC team often focus on assessing the risk factors before committing their capital. The chances of an early-stage company receiving funds from a VC is typically very low as it has to demonstrate serious customer traction in terms of sales, revenue, or engagement. They understand that it is high risk, and while the reward can be significant, they have often experienced many new companies that never meet their targets and go bust. Very few of their investments end up making it to a successful exit event with meaningful returns. They are careful in making their selection, and it can take several months (three to six) to close a round on a particular venture. Thus, you need to be sure to have sufficient funds to survive the period during the vetting and closing process. Some examples of well-known VCs are Sequoia Capital, Bessemer Venture Partners,

Accel, and Benchmark. There are many more that can easily be researched online to better understand their focus industries, size, the funds they offer, and the typical company profile that they invest in.

Below, I shed some light on a few of the main factors that VCs seriously consider when evaluating a company or an investment opportunity:

1. The team. They realize that, no matter how great the idea is, it is the team who will execute the business plan and strategy. Investors seek to build trust and confidence in the team first before making an investment decision. They look for experience, trust, maturity, and, most importantly, a passion to win. They also need to be sure they can communicate with the leadership. Therefore, they will pay attention to communication, the team's responses, and how easy it is to cooperate with the company's executives.

2. The business stage. VCs often choose to invest in businesses that are at a stage where they can demonstrate market traction, thus only considering mature companies. However, some do choose to invest in early-stage companies, where the risk is higher but also the reward can be greater. Early stage does not necessarily mean idea or concept stage; it means that, at a minimum, a prototype is developed, a demo can be presented, or an MVP was launched and tested with a set of friendly users. Some VCs go as far as asking for demonstrating a net annual earning level of $1 million to $4 million before agreeing to review an opportunity.

3. Scalability. VCs look for opportunities that can scale to a significant level so that it can provide a 10- to100-times potential return. Yes, those multiples sound aggressive goals by any measure, but you also need to understand that not all opportunities will succeed, and those that do will need to make up for the failures of others they have invested money in and lost. They would like to see potential for big growth. A market's size is typically measured in billions of dollars and a business may need to demonstrate the opportunity to reach hundreds of millions to billions of dollars in value within a five-year timeline, by capturing a measurable stake in that market.

4. Innovation and differentiation. VCs look for companies that have demonstrated innovation in their capabilities and products and which can demand strong competitive advantage in the market. They understand that competitive advantage can help gain market share. It also needs to be a defendable differentiation where it is protected by a patent, or some other means such as a strong brand or license. It must be complex and difficult to replicate, or provide another defendable, unique edge.

5. Exit strategy. VCs look to recoup their investments within three to five years, but some tend to invest for longer periods. At the end of the period, they look for an exit event. The exit event can be in the form of an acquisition by another strategic buyer or going public in an initial public offering (IPO). Most VCs are not interested in a family-style investment where the company continues to operate indefinitely without an exit event.

6. Investment amount. While there are a few VCs that are willing to invest in relatively small amounts (measured in $100,000s), they are typically interested in investing in the millions of dollars,

175

such as $2 million to $8 million in the first round (also referred to as Series A funding), followed by higher second or third round amounts (Series B, C, D…) where each round can vary in size but some can reach hundreds of millions of dollars. Since they pay attention to each investment equally and they perform the same amount of work in vetting a single opportunity, they prefer to direct their efforts to larger investments as it's more cost effective. VCs often invest collectively in groups where multiple VCs are involved in a single round. There is typically a leading VC in a round, and others join the opportunity as each can participate in the due diligence process.

7. Opportunity fit within their investments' categories. VCs often develop expertise in certain industries and try to group their investments in this way. For example, some focus on business-to-business startups, while others may focus on consumer-based companies, retail, cyber security, big data, social networks, telecoms, among other industries. When a VC chooses an opportunity to invest within their category, the VC team also look to get involved and offer their experience to the company receiving the investments. Many VCs request a chair on the board of directors as a condition, which allows them to get more involved, provide guidance when they can, and stay close to the investment.

Strategic Investments by Corporations (or Corporate Venture Capital)

Some corporations may be strategically interested in the success of an innovative business and may choose to extend an investment offer to support the company in meeting a specific market milestone or succeeding in a customer trial. They may have an interest in using the product that is being developed or they may be interested in partnering with the company in order to create a joint offering to the marketplace and selling it to their existing customers. They may also be considering a future acquisition of the startup. In these cases, they are interested in companies that have common synergies in terms of the industry they are in or wish to expand into, whether the products or services complement their portfolio, or if their offering may become more competitive in the market if they join together. Therefore, one of the important advantages of this sort of funding, is that they can be one of the early customers too. Citi Ventures, Verizon Ventures, Intel, Microsoft, Pfizer, Walmart, and GE, and many others, have a dedicated venture capital organization that focuses on investing in new innovations that are relevant to their businesses.

Preparing for Funding Events

The best funding source for a business is the one that comes from its customers, where the business can generate sufficient capital to pay for its operations' expansion and growth itself. For many, however, external funding is the only option and is required to meet necessary market milestones and to scale at a competitive pace. When assessing the company's funding needs, be sure to know your finances well and understand the business's needs. Preparing the company's records, the team, and the pitch are important steps in order to shorten the time needed to secure the necessary funding.

Not every business needs to go through external funding. Those which do should determine which sources are suitable for them based on their goals and the stage they are at.

LEADER'S SUCCESS QUESTIONS

Principle 10: Secure Your Funding
What Capital Do You Need and How Will You Find Investors?

Ask yourself the following questions to ensure you are prepared when reaching out for external funding.

1. How are you going to fund the business's initial launch and its operational expenses until it can stand on its own?

2. How much money does the business need to become customer ready or reach its first milestone?

3. How much incremental funding would be needed over the course of the next six months to a year, or until reaching the breakeven point milestone?

4. When do you need to raise funding? Understand your timeline in order to start early enough in the process. The funding process can take significant effort, time, and several cycles, and it's a strain on the leader and their team. You must have the necessary capital to operate during the time you are seeking external funding.

5. How much capital have the leaders and/or other investors paid into the company already?

6. Do you have a valuation for the company? If you decided to invite investors to join, could the value be shared with them? Can you justify the determined valuation with hard facts, such as sales, revenue, growth trends, customers registration, or other market data?

7. Could friends and family invest? Consider them first. However, do not borrow an amount that can alter the lives of those who are close to you if the risk is high.

8. Do you need the funds to support the delivery of a guaranteed contract? If so, a bank loan may be a good avenue to help you execute the contract instead of selling equity. Banks are more willing to offer a loan when there is a guaranteed income.

9. Could you approach angel investors? They can provide the initial needed funds for early-stage activities and the process can help you as the founder to learn more about the perceived risk factors through the eyes of unbiased and experienced investors.

10. Are there crowdfunding platforms that would suit your product or business? For some, these may offer the kick start they need. Not all platforms are equal, and some serve different purposes. Be sure you are comfortable with the fees and terms associated with platform usage.

11. Can you consider incubators or accelerators? They can offer valuable connections to potential investors, partners, and customers for friendly trials. They can also be a helpful bridge to larger funding institutions. But be sure to carefully select who to engage by researching their performance and talking with them directly and with others that have engaged with them.

12. Are there venture capitalists and corporate institutions which may want to invest in your business? If your company has demonstrated market success and customer traction and still requires funds to meet additional milestones, then a VC can be a good option. Remember that they invest not only in products, but also in the team.

Principle 11

Prepare Your Pitch:

How Can You Convince Investors, Partners, and Customers to Get on Board?

When meeting with a potential customer, partner, or investor, having a simple and accurate description of your business is extremely important as it represents your ability to articulate your value and services clearly. There are several presentation types that can be used, each one being customized to meet the occasion and purpose of the discussion. This may also include a call to action, if this is the aim of the pitch. The key three types of pitch are as follows:

1. The five-second pitch
2. The elevator pitch (30 to 90 seconds)
3. The detailed pitch (five to 15 minutes or longer).

Each pitching style has a specific goal but all three should be presented using a clear, concise, and attractive approach that delivers an accurate view of the company and generates the desired impression and interest from the audience. This is often easier said than done. Pitching is an art. However, if it's done right, it can make a great difference in securing the resources you want, the funding you need, and landing the key customers for your venture. Preparation, fine tuning, and practice is the key to mastering the delivery of the perfect pitch. Soliciting feedback from a friendly audience is encouraged as it can provide valuable input from an objective party.

The Five-Second Pitch – A Short Value Proposition Statement

The five-second pitch is a short value proposition statement. It is intended to convey the company's value and be understood in a simple, yet well-crafted, statement. Many businesses have a complex value offering, delivery model, or functionality. However, that does not necessarily mean it cannot be expressed in a simple form that can be understood by the intended audience. Simplifying the value proposition is a challenge, but you must work at it iteratively so that anyone who hears it for the first time, regardless of their background, can understand and develop an appreciation of the product. The five-second pitch includes only the "what" statement, or the value being offered, and should not include "how" the value is created. This is not the time to talk about the mission, or vision statement; this is simply one or two lines describing what the business is about, what it offers, its value, and who would benefit from it.

It might be used in a discussion when someone asks what your business does or what it is about. It can ignite interest, and allow the audience to be intrigued to ask and learn more. For example, I might describe the Google search engine as follows:

"A web application that allows users to search for, and find, relevant results for a particular keyword or keywords on the world wide web."

A simpler form might be:

"A web application that helps users search for anything online."

Let's take another example, perhaps a venture that is building residential rental properties focused on college students. It can be described as follows:

"We build multitenant residential rental properties for college students."

It is the shortest possible description of your business that can define the value you offer and the audience you serve.

The Elevator Pitch

The name says it all. The "elevator pitch" is a short description of the business that can be expressed in the time spent in a typical elevator ride; 30 to 90 seconds. In a pitch like this, you are expected to elaborate on the business to induce enough interest from the audience that they want to learn more. Whether it is a pitch to potential investors, partners, or customers, the elevator pitch must include the "what" i.e., the value you are offering, the "who" i.e., who benefits from it, and anything unique that differentiates your business, including a short description of the offering itself, and whether it is a product or a service. It needs to be clear, simple to understand, well presented, and prepared ahead of time. One of the key elements of a successful elevator pitch is the readiness of the material and your level of practice; this ensures the pitch can be structured and finessed to serve its intent and objectives.

There are commonly two types of elevator pitch. A sales pitch and an investor pitch. A sales elevator pitch is intended to convey the value to the customer and set the company apart from competitors. The pitch to a prospective customer would include "what" the offering is, the benefit to the customer, i.e. why they should care, and how unique it is compared to others. This can include any significant successes you've had, and a call to action i.e., offering a demo, trial, or another meeting to discuss the unique solutions you could offer to solve the customer's problem or help them to capitalize on the opportunity.

A pitch to investors, in contrast, is more focused on the growth opportunity and the value to investors who may be deciding whether to put their own capital into the product. The pitch to investors would include not only the "what" and its benefits to customers, but also the potential growth opportunity by conveying the market size, and the success you may have had in the market in generating revenues, sales, or customer traction. The call to action would be to ask them if they would be interested in a meeting to discuss the details of the company and learn more about the opportunity to invest.

In both of these, you may want to consider leading with a short introduction outlining who you are and describing the company, depending on how familiar the audience is with you. Some presenters choose to wait and not introduce themselves right away, as they may feel they can add an innovative and persuasive opening to attract the audience's attention rather than just stating their name.

Elevator Pitch Example

For example, you may start by saying,

"Hi, I am John Smith, the head of product at Company Cool. Every year, more than 90% of companies face the problem of spending too much money on developing low-value products. This results in a significant loss of opportunity, lower capital efficiency, and poor customer experience *(stating the problem, why it is urgent, and why they should care)*. To address this problem, we have developed a unique solution called "Hero," an artificial intelligence-based application. Once it's integrated within your processes, it has proven to reduce or eliminate this problem while increasing your bottom-line profitability by up to 35% *(stating the solution and its value)*. Unlike other solutions in the market, you do not need to change your infrastructure, or location. We can accommodate your current company's structure and solve the problem in half the time and at half the cost when compared to other solutions *(stating the competitive advantage)*. We would love the opportunity to sit with you to discuss how best you can experience and deploy this solution in your business. *(call to action)*."

For an investor pitch, I would replace the last sentence with the below:

"The market size for this opportunity is approximately $50 billion worldwide, and we are uniquely positioned to capture a healthy market share, leveraging our proven differentiated capabilities. Our revenues *(or replace with "our sales," "our customer engagement," or any other demonstratable measure)* for the past 12 months were $2 million and they are growing at the rate of 20%, month over month. We are inviting serious investors to be part of our growth story and would love to talk with you to discuss your potential participation in our success."

In stating the problem, you may add any credible source that can confirm the pain point, such as: "According to X institution, this is a major problem costing the world $X billion, including loss of opportunities, increased stress, and impaired health," or any relevant measure to your business or the opportunity.

There are quite a few resources to help you craft your business pitch and you should utilize as many of these as possible.

The Detailed Pitch

Now that you've developed your elevator pitch, a more elaborate and detailed pitch is needed to support those who want to dig deeper into your business and learn more about the opportunity to engage with you. The five to 15-minute detailed pitch should take the audience on a journey and keep them interested and engaged. The pitch content should include many, if not all, of the elements discussed in your business plan and go-to-market strategy, with the exception of confidential material or information that's not relevant to the audience.

Think about the order and flow carefully. A good presentation should start with an exciting statement or an engaging call to action. Many successful presentations start with a short, interesting story to draw listeners in. People are much more attentive and focused when listening to a story than when being lectured. Another approach is starting with an engaging question, such as how many of them have been in a given situation or seen a specific event that can be tied to the presentation. Asking a question helps the audience to relate to you or the discussion you are about to start. Once you pass the initial engaging statement, story, or question, you may introduce yourself and proceed with the prepared presentation.

You must also consider your slides and how to use them persuasively, rather than drown your audience in information. There is a difference between a set of slides that can stand on its own, covering all of the material in detail, and slides that are used creatively to support a pitch. You may start with a picture that represents the story you are starting with, or a simple statement that represents the introductory question. Not every word you are about to speak needs to be included. Each slide only needs to be engaging with one to three main points. As a presenter, you can elaborate on each point. However, collateral material that is shared as a reference point with the customer or investor, such as a product documents, brochure, or company's offer, can have more detail.

Example of a Detailed Investor Pitch

Below is an example of a good flow for an investor pitch with the purpose of generating interest to invest funds in exchange for a stake in the company.

1. The team. Present the team and highlight their key strengths. Establish trust and confidence in their capabilities, skills, and experience related to the business. The audience will also be gaining insight into how well the team can communicate and how easy it is to work with them in the future.

2. The problem. Describe the pain point or the problem that the business has identified; a need to be solved and addressed. This should include why it's important to solve it and how urgent it is.

3. The customer. Discuss who the target customer segments are and why they will care about the product. How are they addressing the problem today without the product and why are existing solutions not fulfilling this need? Support the discussion with facts and data.

4. The mission. Describe the mission statement and what your company wants to do about the problem.

5. The solution. Present your solution, product, or service that addresses this opportunity. This should include a brief description of its components and how it works.

6. Differentiation. What makes it unique and different from other products in the market? Why would someone choose your solution over its competitors?

7. Competition. Who are the competitors and how do you stack up against them? What are their strengths and weaknesses in comparison to yours? You may choose to perform a SWOT analysis or a similar tabular matrix.

8. Finances. Demonstrate your knowledge by presenting your key numbers. Describe your financials, including pricing, cost, the opportunity size over the next one to three years, number of customers, and associated revenue you are capturing or want to capture. You may want to add a longer projection period such as five years. The key here is to be able to describe the projections based on reasonable and believable assumptions. Use a bottom-up approach that demonstrates customer growth month over month, units sold with prices, and expenses on a monthly or quarterly basis. This is more accurate than a top-down model where you assume a percentage gain from the overall market size; this simply demonstrates the opportunity potential.

9. Go-to-market strategy. Describe your sales and go-to-market strategy. Discuss how you are going to achieve the projected finances in order to make those projections achievable and credible.

10. Funding. Describe the level of funding you need, what has been received so far, if any, from previous funding events or rounds, including any self-funding. Discuss what you need moving forward, the instrument of finance that you are proposing, such as straight capital in exchange for equity, convertible notes, a loan, etc. Outline what the funds are going to be used for, and what milestones you will hit by what timeframe. For example, funds might be used for a market trial, to grow the MVP, to deploy the product in a particular city or country, to expand a market, to hire skilled employees in a specific capacity, to grow revenue, or to build and expand your competitive advantage.

You may not want to list all the details of your plans, but the intention here is to develop trust and confidence in your team, your products, and the business growth potential for the investor, so they can see the opportunity and have a robust discussion. They may offer to invest if they can see a way to make money, or they may say no to the opportunity because they see too many red flags. They may pass on the investment because the risk level is too high due to existing strong competition; because the business does not have defendable differentiation; because the team's skills are not sufficient to navigate potential challenges; because it's too early to invest and the business has not yet demonstrated sufficient success; or if it is out of their industry's scope of interest. If you are rejected, try to find out why so that you may address these concerns in your next round of discussions with other future VCs. It may be that you need to optimize your pitch or wait a little longer until after you have mitigated the risks by advancing your venture in the market. I have met entrepreneurs who, when asked about their initial funding needs, have requested a substantial amount such as $10 million or more. This really reflects the needs of a full operating company when, in reality, most ventures can get started and reach demonstratable market traction, produce an initial prototype or an MVP, with much less (often under $1 million or even under $100,000).

Example of a Detailed Sales Pitch with a Potential Business Customer

Always be prepared. Learn about the customer, their business, and their problems ahead of time. Attempt to identify their needs and how your solution or offering may solve those needs. When you meet, listen to the customer. You may start the meeting by asking them, before you give your pitch, if they're willing to discuss their current priorities and the pain points they're facing. The research you did before meeting can help significantly, including identifying the key team members that should be in the meeting, including decision makers.

The presentation should start by establishing credibility in your business and product and then moving on to discuss the solution and how it addresses the customer's needs. An example flow is as follows:

1. **Company.** Describe the company's history, strengths, and existing customers. Establish credibility and trust in the company by expressing your understanding of the market, the industry, and outlining your success stories so far.

2. **State the problem.** Describe the opportunity or the pain points as you see them in the market and confirm if the customer is experiencing the same problems. Quantify the problem if possible and how impactful it is on them or their customers, perhaps by discussing revenue or any other metrics important to them. While the solution you intend to present may address or solve many problem types, be sure you state the problem that is specifically related to the customer, and not in general.

3. **The mission.** Describe the mission statement and how this captures what your company is doing about the problem.

4. **The solution.** Present your solution, product, or service that addresses this opportunity. Give a brief description of its components, its architecture if applicable, and how it works. Discuss how it is implemented and used by the customer to resolve the pain point, or how it offers value and benefits.

5. **Implementation.** Discuss what it would take for this product to be acquired by the customer. Highlight the process to access the offering. It may be as simple as setting up online access, or it may require physical installation and integration.

6. Demonstration (demo). If possible, provide a real-time demonstration of the product, how it's used, and highlight the ease of use, features, and benefits. This is a key step where you have the chance to showcase your offering and impress the customer.

7. Differentiation. Discuss what makes it unique and different. Explain not only your competitive advantage, but also how the customer will gain a unique edge if they use your offering.

8. Advantages and market results. Discuss the advantages it provides and present real data if available. You may use data from friendly customers who used your offering and provided you with permission to share. Data from previous trials with users may also be used.

9. Pricing. Depending on the maturity of the relationship, the stage of the sales process, and the type of offering, you may choose to present the pricing structure, or pricing options, especially if you feel that your pricing strategy is an added competitive advantage.

10. Feedback and follow ups. Ask for customer feedback and propose a call to action follow up. Some customers may offer feedback throughout the presentation, while others may choose to wait until the end. Allow time for questions and clarifications. Ideally, you may arrange a call to action, such as a follow up with the same team, or other key members that are not present. Try to learn who the decision makers and influencers are and what their typical purchasing process is. Based on their feedback and input, you can structure these next steps to advance the sales process.

When creating your pitch, remember to distinguish between the three types of pitches discussed above and understand the goal that you are trying to achieve from each. Prepare each according to the setting, the audience, and the stage the company is at.

LEADER'S SUCCESS QUESTIONS

Principle 11: Prepare Your Pitch
How Can You Convince Investors, Partners, and Customers to Get on Board?

Ask yourself if you have these pitches below ready. Have you already tested them with friendly audiences, such as team members in your company and advisors or mentors, before delivering them to external customers, investors, or the public in general?

1. What is your five-second pitch? (one or two sentences outlining your value proposition)

2. What is your elevator pitch? (30 – 90 seconds)

3. What is your detailed pitch? (five – 15 minutes)
   a. To customers
   b. To investors

Principle 12

Evaluate Your Risk:

What Is the Likelihood of Failure and How Will You Prepare for and Mitigate Your Business Risk?

Evaluating risk is not about discouraging yourself or introducing friction to the launch process of the business. Rather, it is a way to make sure there is an unbiased assessment of what can go wrong, how that could impact your business reaching its goals, and how you can prepare appropriately for it. Once you have assessed what could go wrong, you should also assess how likely it is that these risks will occur and their potential level of impact. Of course, many things can go wrong, but not all risks are equal.

This step is possibly one of the most challenging when you are developing a business. When one is eager and excited to launch a new venture, our optimism takes over our emotions and can easily blind us from seeing potential risks that can be easily addressed if recognized in a timely manner. Not addressing risks early in the process can severely reduce the possibility of reaching your targets. At this stage, I would encourage the leader to start with a general brainstorming session and list all possible risks that may face the business. Once compiled, each can be evaluated and ordered by priority level to determine which risks have measurable impact and need to be addressed, and which have less significance to the venture's success and therefore can be parked or dealt with at a later time.

To prompt a risk assessment brainstorm, start by asking yourself the following questions:

1. What can make the business fail? Identify the top three to five factors.
2. What factors will make your business succeed? Are they likely to happen and why are they so important?
3. Have others with a similar business concept tried to launch before and failed? Why do you think they failed?
4. Have others tried it before and succeeded? Why do you think they succeeded?
5. How confident are you with the research done in verifying the business assumptions?

Analyzing and Mitigating Your Business Risk

Business risks can be generated through one or multiple factors and are often the result of changes in market conditions, rising competition, low customer demand, and internal failures in the organization. The impact of those risks can be felt directly on business performance if they are not addressed, so it's important to analyze them and plan ahead of time your approach to mitigating them.

Ask yourself the following questions to help prepare for the worst and thus increase your chances of success:

What If Customers Don't Come and Don't Need My Offering?

This is the first question you should ask yourself and possibly the biggest risk to a business. Examining its probability, and why it may happen, is vital. If your customers don't come at all or trickle through at a much slower pace than projected, this could have a devastating impact! It could be due to the offering missing its target objective and not delivering the intended value. It may be due to overestimating demand because of lack of sufficient research or biased and weak analysis. The cost of customer acquisition may also prove to be much higher than planned, and the sales cycle may be longer and more complex than anticipated. All these factors can put significant stress on the business and make it harder to navigate out of the storm. Therefore, it's critical to confirm, early on, the value of your product with customers, the urgency of the problem being addressed, the likely demand, and the cost and time to acquire customers. This includes assessing the competition's strengths and their impact on customers' decisions to buy from your business.

An honest assessment of what can go wrong helps you reexamine your assumptions and solidify your business plan. There is a well-known saying in business: "If you wait until your product or service is perfect to launch, then you are too late." The point is to emphasize the benefit of launching fast, learning from your mistakes, and addressing them in future iterations. If you wait too long, it is likely the market will have shifted by then, or competitors may have moved faster than you and gained a bigger market from your share. I agree with this approach in general, but it's still absolutely essential to develop the core features of the business before progressing to market. Perfection can wait, but the minimum requirements to meet the value proposition of your product must be in place before launch.

The problem is that some can easily confuse this advice to get to market quickly with accepting any level of risk, regardless of impact level, and a belief that they can just fix the identified problems later. There is a big difference between going to market with an imperfect product and having missed the mark entirely on the minimum required elements to meet customer expectations. Google, which is known to be one of the pioneering companies in building and introducing innovative products and services that have truly revolutionized our lives, has not always been equally successful in introducing all its products. For example, they launched Google Glass in 2014, a hi-tech wearable product that was introduced with significant hype. Users would wear them as they would normal glasses, yet they could do voice-activated searches, follow directions on a map, and take pictures, among many other amazing features. However, the product has failed to gain traction after several missteps during its launch and, while there are many reasons that have contributed to its failure, not receiving a positive customer response is possibly the main one. The price was considered expensive at $1,500. The design seems to have been rushed without sufficient validation from all target segments, as it wasn't seen as appealing or stylish. There were also too many software bugs and complaints that were raised by customers that possibly should have been resolved before going to market. But one of the major issues was the serious privacy and safety concerns around the collected data when using the device while outside in public or in a private setting. Such issues could have been identified as risk items and addressed through the product design, features, and marketing messaging, which may have led to a different outcome. The product was discontinued in 2015 after multiple attempts to revive it.

In addition to functionality and product quality, pricing structure, buying process, location, and other factors may also be the reason why customers do not show up. Examine all relevant factors that might impact your customers decision to buy, and prioritize based on their importance in order to address those that can make a real difference before going to market.

What If Customers Find My Product Too Expensive?

If your customers find your offering too expensive or the sales model unattractive, then adjustments will be needed to address this risk. Know your finances well so as to be prepared for optimizing your pricing and sales structure if needed. Understand your flexibility in driving your cost or your margins down. You may want to consider updating your marketing message and channels to highlight and clarify the value of your product and to ensure your customer develops a clear understanding of why you've priced the product at a given point. You do not need to be the cheapest provider in the market, but you do need to justify a premium price to customers based on the value offered. With that said, if your product's pricing is impacted by competition and market driven, you must be sensitive to this context and price accordingly, if margins allow. Knowing your pricing limits can save you valuable time when you need to adjust your price.

What If New Competitors Emerge Within a Short Period of Launch, or Existing Competitors Evolve to Present a Similar Value Proposition?

There is always the risk of new competitors appearing in the market with a similar value proposition or existing companies developing offerings that are in direct competition with yours. While this can certainly impact your market projections and sales, it is somewhat reassuring that your offering is attractive enough to spark competition and has demand and market interest. In this case, you would need to analyze your competitive advantage, your strengths over the competition, and how well you can stand up to newcomers that may emerge. For example, you may want to compare the quality of the product, the embedded functionalities or features, the speed of delivery, the accessibility of the product, its durability, your customer service, and pricing structure. Make sure you fully understand the competitors' strengths and weaknesses in order to take action when necessary in addressing this risk.

Will the Business Have Sufficient Funds to Launch and Operate Until it Can Be Fully Self-sufficient?

Funding availability and raising new funds is a risk that a company must plan for. It can take several months, and sometimes longer, to secure a meaningful funding source. Have a plan ahead of time and start early in securing the capital needed for the business. How much runway do you have? What if the breakeven point takes longer than planned? Can there be steps taken to accelerate the arrival of the breakeven point to de-risk funding issues? The answer could be in starting the planning and execution efforts early. You may also need to identify multiple funding sources that you can take advantage of as the need arises. Raising *more* funds early on in the process, in order to have additional contingency funds, can also be an option. For example, some opt for raising 25% to 50% more capital than their plan requires in order to address capital and market risks. Plan your financial commitments to the vendor carefully as this can also work to reduce your risk. For example, some companies negotiate long-term payment plans with their vendors that can coincide with certain milestones, such as timing of expected revenue streams, contract execution, or closing of a funding round.

Are There Measures in Place to Track Progress?

A key element of preparing and planning for risk and failure is understanding what it means to not meet your intended goals. For that, indicators and measures can be developed to track how the business is progressing and to indicate when the business is derailed from its intended path. The measures can include how closely the business is meeting the original sales projections, or customer numbers by a certain date. Other measures could be operational, such as how successful the business is in acquiring the needed resources, or if it is meeting specific milestones within the plan timeline. Having a set of indicators to measure and track success can help its leaders to assess the health of its operations day by day and determine how to continue to grow, or to pivot and optimize as necessary. Examples of measures could be sales and revenue growth rate month over month, customer registrations, number of users within a certain period, number of visits, unique visits and revisits, or customer feedback and ratings.

These indicators are also important in guiding stakeholders, in the unfortunate scenario of critical failure, as to when to stop and pull out. Yes, it is a risk and ceasing operations can sometimes be the only option. Understanding when to do this and what indicators can trigger such a conclusion helps to reduce unnecessary losses and save precious time that can be used toward future opportunities.

What Can Be Learned from Other Companies Launching Similar Products?

First movers have great advantages: they are first to market in establishing a strong brand for a specific value offering; they enjoy a low level of competition; and they can develop customer loyalty where the cost of switching to future competitors may be high. However, it is often the second, third, or late movers that take the leadership position in a given business category. The reason is that a second mover can learn from the mistakes of the first mover and study customer response as they modify their product functionalities and market strategy to gain traction. Apple is a prime example of a company that has become a leader in every product or service category they operate in, yet they have rarely been a first mover. For example, they did not make the first personal computer, mobile phone, music player, digital watch, or tablet. Yet, in every one of their products or services, they leveraged the significant market data that existed from previous companies and they learned from it. Once this data is well understood, and when coupled with their creative process focused on exceeding customers' expectations, they produce superior products and services that are often unmatched by any current or previous competitors.

Examining your plans and comparing with what others have done can be very helpful and provide great insight. When asking why similar businesses succeeded and why others have failed, or fell short, the answers will provide a tremendous advantage and save you capital by avoiding others' mistakes and learning how best to meet or exceed customer expectations.

Factors That May Cause a Business to Fail

Factors that can cause a business to fail do vary. It can be a single factor or a combination of many. The list below is not comprehensive, and some of these factors may have different degrees of impact than others depending on the business and the industry. However, going through them can help you to assess your readiness for market, and prepare your business to address them.

1. **Not having a plan.** This is a dangerous approach and a business must not proceed without one. Use the principles discussed here to develop one.

2. **Misreading the market's needs.** Offering products or services that are not urgently needed by your target customers will mean very low demand.

3. **Insufficient research** in identifying the opportunity, the market size, and competition.

4. **Team related risks:**
   i. Bad execution of a good plan. You must strengthen your team and enhance your skills to avoid this.

ii. Inexperience of the team and a lack of skilled resources. Hire carefully and according to proven skills aligned with your business needs.

5. Inflexible practices. Lack of adaptation to new discoveries and market changes. Take market signals seriously and take action early to avoid becoming obsolete. Nokia's failure with their mobile phones is an example of an inflexible strategy and a lack of adaptation to clear market signals.

6. Lake of sufficient funding. Plan ahead and always assume you will need an extra 25% to 50% contingency when starting.

7. Bad business culture. An uncooperative, uncollaborative culture that is not customer centric can lead to a fragmented organization with a lack of focus and loss of business.

8. Lack of measurement of progress. There must be key indicators to assess successes, declines, losses, challenges, and customer satisfaction. You need to understand what you are doing well, what you are doing badly, and what needs to change. Develop measures to keep your risk assessment timely and which allow you to respond to market changes.

9. Product issues. Be wary of:
   i. functionality that does not meet customer expectation;
   ii. poor product quality compared to the competition;
   iii. usability that is not customer-friendly or engaging.

10. **Unsuccessful sales model:** For example, your product may not be easily accessible for customers; the sales process may have a lot of friction; the product may be difficult to discover, buy, have delivered, or receive support for.

11. **Incorrect pricing or pricing structure.** Pricing must be competitive and compatible with customer affordability.

12. **Lack of sufficient competitive differentiation.** If your competitors develop a better offering or substitute. and you do not adapt or optimize in response, you will lose your unique edge.

13. **Unresponsive customer service.**

14. **Faulty financial projections.** Underestimating the costs, or overestimating your revenues, will skew your financial projections.

15. **Poorly thought-out business strategy.** For example, the wrong go-to-market rollout strategy, sales strategy, or your plan of how to reach customers.

16. **An inability to obtain required licensing or receive a required patent.**

17. **Lack of internal controls on financial performance, quality of production, and the efficiency of operations.**

18. **Complex internal operational processes and unnecessary bureaucracy.** This can cause inflexibility and an inability to innovate and adapt to opportunities.

19. **Natural disasters, wars, and major barriers developed by government.** This can lead to a challenging business environment and, ultimately, failure. Although they could be mitigated or prepared for, these are the most difficult risks to fully control.

It is impossible to address all risk items before launching and, as discussed earlier, it may not even be good practice to do so as not all issues are equal in importance to the early success of your business. Proper risk assessment in identifying the priority list and addressing them ahead of time will help the leader to strategically develop a defensible position, determine what must be in place upon launch, and what can wait and be addressed as part of your future roadmap.

LEADER'S SUCCESS QUESTIONS

Principle 12: Evaluate Your Risk
What is the Likelihood of Failure and How Will You Prepare for and Mitigate Your Business Risk?

Ask yourself the following questions to help prepare for the worst and thus increase your chances of success. What can make my business fail? Identify the top three to five factors and have a plan to mitigate them.

1. What if customers don't come and don't need my offering? Don't miss the mark on this one. Validate their interest and the value of your offering before you go deep into your new venture. Go back to Principle 1.

2. What if customers find my product too expensive? Examine your messaging, your value proposition, and differentiation. Research your competition's pricing and understand your cost structure to determine your pricing flexibility in offering new models that are more acceptable to customers.

3. What if new competitors emerge within a short period of launch, or existing competitors evolve to present a similar value proposition? Be prepared for the financial impact and add a risk margin to your revenue projections. Be ready to compete in differentiating your business, and be prepared to grow your market advantage through your product's roadmap, marketing, and in your passion to satisfy your customers.

4. Will the business have sufficient funds to launch and operate until it can be fully self-sufficient? Understand your capital needs and don't assume the best-case scenario. Have a plan to beat your market goals, but be prepared for falling short. Identify capital sources that you can utilize, and understand the time required to access it should you need it.

5. Are there measures in place to track progress? Understand your performance against important financial and operational measures to evaluate the health of your business, the level of risk, and how close you are to meeting your goals. When you recognize a deviation in its early stages, quick action can be taken to correct.

Part Five

Time for Action

Ready to Take Action? Let's Build Your Plan!

Next Steps?

You have the idea, you've identified the opportunity, and you are ready to act. This is an exciting phase and you absolutely should be raring to go. If you believe it is time to start, then do it. Go and build your product, service, or company. But first, before you go any further, build your plan.

Building the plan will enable you to understand the scale of the journey you are about to embark on, its potential rewards, and the risks associated with it. Choosing when to begin must be balanced with your current responsibilities and financial needs. For entrepreneurs, you must recognize that your obligations in life do not stop when you're launching your new product or company. You will still have bills to pay and, if married with a family, you may have a mortgage, kids to feed, school fees, or college funds to build. You may already have your dream job, giving you a bright future, which is hard to give up. For those who have multiple sources of income, either within the family or as an individual, the risk is possibly more manageable.

For corporate leaders, while the risks may still be high but usually less catastrophic, your capital is not limitless, and you must choose where to spend your time and money in launching the next opportunity. The success of a new product will have a significant impact on the company's success, your own career development, and that of your colleagues. Therefore, developing a sound plan is critically important to gain the trust of the leadership and your managers, who will need to be on board in order to proceed.

Following these principles is fundamental to the success of developing and launching a new business or product, and it will maximize the effectiveness of the capital, time spent, the chances of meeting financial goals, and its overall sustainability.

If you are ready, then the next step is to capture the 12 Principles in a plan of your own that outlines your strategic business decisions. The plan does not have to be completed in one sitting and isn't set in stone – it can evolve and progress over time. Each of the principles should be considered and there should be a contribution from the leader, the team involved, advisors, and even trusted third parties who can offer insight.

Research is an essential component in building a sound strategy and informed assumptions based on up-to-date and detailed knowledge. Customer studies, market analysis, and product analysis are all helpful and typically considered a prerequisite before finalizing your plan. Do your homework.

Identifying what you know, and what you don't, enables you to fill gaps in your knowledge. Start by going through each of the presented questions and drafting a response based on your current experience, understanding of the situation, and the products you are being asked to plan for. Then, leverage the resources that you have access to, such as online materials, trusted advisors, or friends who are familiar with the subject being worked on. During the draft for each response, mark the questions that cannot be answered immediately. Don't get stuck on a single item or those that require further research. Complete the first draft, and then review the areas that require additional answers or further input, and use this to shape your research.

The 12 Principles Business Plan Template

Below is a template that brings together the essential questions and guidance from this book in an easy-to-use framework for building your own business plan. You can modify and adjust as needed in order to create a bespoke guide that will help you to define, validate, fund, and pitch your idea successfully. Why fail when you can win?

1. Define Your Opportunity: What Problem Are You Trying to Solve?

Discuss the problem you are solving and establish the pain point/s the customer is facing.

   a. What problem or need you are addressing?
   b. How urgent and critical is this problem to your customers? Do you think they will perceive your business, product, or service as important to them and worth prioritizing?
   c. What are customers doing now to address the problem you have identified? Why do you think a new solution is necessary?
   d. What customer behaviors (operations, processes, or way of living) are you seeking to change or introduce? How are you changing these behaviors with your offering and how does that compare to the behavior needed to use existing solutions?

2. Understand Your Customer: Who Are They and How Many Are There?

Ask yourself the following questions to help you explore these issues in the customer assessment section of your business plan:

a. Who are you trying to help? Who may want to use the business's offering? As applicable, specify the target segments such as age group, education, locations, interests, income level, and career type. If it is a business-to-business offering, then describe the industry, the functions, user skills, and role in the company. You should take the same approach if it is targeting the public sector or a government sector.

b. What is the initial customer segment that you want to target and acquire?

c. How large is the market? How big is the opportunity within the market? This describes the total market size and the potential market share which you can obtain.

d. Is the target industry growing, stable, or declining? Are customers growing in this segment or declining?

3. State Your Business Mission: What Are Your Values and Goals?

Ask yourself the following questions to help craft your mission statement:

a. What is the core passion that you want to commit to addressing for your customer?

b. Who is the target customer or audience that will benefit from your efforts?

c. What is the solution that you will offer to fulfill your commitment to customers? Examine your mission statement and be sure you can set goals and objectives that you can feasibly achieve. Once you've established your mission statement, then define your objectives:

d. What are your business objectives and goals? Define the goals that, if satisfied, will support you in meeting your mission statement and fulfilling your commitment to customers.

4. Choose Your Team: Who Will Translate Your Mission into Reality?

The leader, the team, and your personal and professional networks are the prime catalysts to translating the business mission and its objectives into a successful reality. Every team member should have a skill set that is relevant and important to their role and to the success of the business. Your secret sauce may be residing here.

Ask yourself the following questions to help you select the team that will translate your mission into a reality:

a. What resources, skills, experience, functions, and roles do you need to launch a successful business?

b. How do you plan to inspire them, to empower them, to reward them, and to develop a collaborative culture to maximize their commitment, effectiveness, and the quality of their deliverables?

c. What key measures do you need to have in place for each role to effectively evaluate each employee's performance?

d. How do you plan to compensate them? Is it pay only, equity only, or a mixture of both? Do you need to offer benefits, and what are the costs of these benefits?

e. What additional resources are needed beyond the core roles? For example, human resources functions and payroll.

f. Is there a need for hiring extra resources from contractors or third-party resources from partners?

5. Specify Your Product: What Are the Key Functional Components, the Requirements and the Long-term Roadmap?

Ask yourself the following questions to help define your product, its MVP, and how it will develop (plan big, but start small and agile):

a. What is your solution?

b. How would you define it at a high level, in plain and simple language which any user can understand? This must reflect the research that helped to define the opportunity, the customer pain point that it addresses, and the target customers.

c. What are the detailed functional components of your business? List and describe each in brief, yet with enough detail so they can be clearly understood. Providing clarity on the functional capabilities will help you and your team to translate those requirements into reality.

d. What is the short-term initial functionality that would be necessary in the minimum viable product (MVP)? How long will it take to develop the MVP? What is the project plan, milestones, and timeline?

e. What is your long-term roadmap? Define the product and functional roadmap for the first six months, and for years one, two, and three. Outline the mature product view and its evolution after the MVP.

f. What would it take to develop it and offer it in the market? This should outline resources, skills, knowhow, and the availability of components, if there is a dependency on a third party.

g. Are there any dependencies on partners or external entities to complete the offering? In some cases, the more partners or dependencies that are required, the more complex, lengthy to build, and expensive the offering can become. Try to understand the potential partnership challenges ahead of time and be prepared to resolve them as they arrive.

h. What measurements of progress and key indicators are in place for tracking business success?

  i. Define milestones with timelines.

  ii. Define customer satisfaction measures.

  iii. Define indicators that allow you to identify how the offering is meeting the desired customer value, including customer engagement.

  iv. Define indicators to measure how the offering is meeting the defined functional requirements.

i. Would you need to file any patents or trademark protection to reserve the rights to any unique functionality, process, design, logo, or service?

6. Identify Your Competition: What Are Their Strengths and Weaknesses and How Do You Compare?

Ask yourself the following questions to help articulate and understand the competitive landscape:

    a. Who else is offering similar value, products, or services that can compete and gain market share from your business?

    b. What are your competitors' strengths and why are customers buying from them?

    c. What are your competitors' weaknesses in terms of meeting customers' expectations?

    d. Which are the top one to three market leaders among competitors and why?

    e. Can you project your market share vs. your competitors? Will that share be sufficient to cover your operations and desired profit margins?

7. Determine Your Competitive Advantage: What is Your Differentiator and Unique Edge?

Ask yourself the following questions to discover your competitive advantage:

    a. Why would customers use your product or service instead of others?

    b. What makes your offering distinctively different, and how valuable is this to the customer?

c. How are you achieving your competitive advantage? Is it tangible through product features, pricing, distribution, time to market, first market mover etc.? Or is it intangible, such as better customer service, a better engagement process, relationships and connections, brand awareness, easier access, a better experience, or free or faster delivery?

d. How does your business stack up against others when performing SWOT analysis? Are you counting on gaining market share from competitors? If so, describe how you will develop a more competitive operating model.

e. Can others easily copy your competitive advantage? How difficult is it for others to acquire the same differentiation? Do you have patent protection granted to your product, design, or process?

f. Do you, or your team, have unique experiences, contacts, and skills that can provide an intangible competitive advantage? Can they support the differentiation of your offering in the market? Your relationships with potential customers, your knowledge in the industry, and your skills in developing a solution can all help give you a unique edge.

8. Know Your Financials: What Are Your Cost, Pricing, and Revenue Structures?

Cost

What is Your Cost Structure for Launching and Operating Your Company?

Ask yourself the following questions to construct an understanding of your cost models:

a. What are the total initial funds required to launch the business, the product, or the service? Your set up costs could include legal costs, incorporation and government registration, licensing and other administrative costs, lease or real estate expenses, equipment, supplies, and furniture.

b. What taxes, interest, and other financial obligations do you have that will impact sales, revenue, and cost projections?

c. Do you plan to pay yourself or cover your living expenses? If you don't have other sources of income, how will you cover your personal obligations?

d. What is the cost of developing the MVP?

e. How much will it cost to test the MVP in the market with customers? Would trials be needed?

f. What are your fixed costs and your variable costs?

g. As applicable, what is the cost of producing a single unit of the product?

h. How can you improve your operational efficiencies over time without compromising the value or quality of your product?

i. How many resources does the business need in order to produce the product? Are they full-time, contractors, or a mix?

j. How long will it take for the business to be revenue ready and sell the first product?

k. How much would it cost to become customer ready, i.e., to engage and sell to customers?

l. What other costs are needed for market readiness? Is there a third-party cost that needs to be considered?

m. What are the marketing costs needed to bring awareness to market?

n. What are the sales costs? How much would it cost to acquire customers and sell the offering (resources, advertising, tools, etc.)?

o. Have you defined the cost to fund the initial launch and then the roadmap over the next one to five years? Segment the cost per milestone that needs to be reached over a certain timeline.

Pricing

What is Your Pricing Structure?

Think of the pricing policy competitively! Then define it! Your price and its structure can be a competitive advantage. Don't think of your profit desires only; think of customers and their willingness to pay or engage with your pricing model.

Ask yourself the following questions to determine your pricing model:

a. What is the pricing structure that is suitable for your business and who pays for the product?

    i. Buy-to-own.

    ii. Lease or subscribe-to-use.

    iii. Third-party sponsored — when a product for consumers is subsidized or paid for by a business customer or governments e.g., advertisers on Google Search enabling the public to use this service for free.

iv. Pay-per-use.

v. Base subscription with pay-as-you-go.

vi. Freemium - where the basic service is free with premium subscription options or upgrades for a fee.

viii. Buy-to-own with an annual subscription fee — such as an initial purchase fee, then a monthly subscription for annual support and maintenance.

b. What is the most appropriate pricing model for determining the price value of your offering? Is it cost-plus, value-based, or market-based?

c. How are your competitors pricing similar products?

Revenue

What is Your Revenue Model and How Will You Track Your Progress?

Explain your method and support your assumptions with credible data. The model should be developed from the bottom up, based on monthly, weekly or daily numbers of units sold and the price per unit, which should provide reasonable assumptions on growth month over month. A top-down projection which assumes a market share as a percentage of the overall market size should only be used to provide a high-level view of the opportunity size; it cannot be used with a high level of confidence for revenue projections.

Ask yourself the following questions in order to develop robust business revenue and value projections:

a. How many customers do you think will use the product per a given period and how often?

b. What are the monthly revenue projections for the first one to three years? Some choose to project five years or more. When developing long-term projections, the business must acknowledge that there will be market fluctuations and unforeseen risks, and therefore projections beyond the first three years should be viewed with lower confidence levels.

c. Are the sales and revenue projections defendable? Why would anyone believe the projections? Do an honest self-assessment of the assumptions and facts used and be sure you convince yourself as a business leader before you try to convince others such as investors or partners.

d. How profitable is the business and when do you reach the breakeven point? What is your desired and projected rate of return? What is your payback period as an associated strategy?

e. What measures do you need to have in place to validate meeting your desired projections? This might include sales, customers, return visits, and other measures that are relevant to your business's revenue.

9. Build Your Go-To-Market Strategy: What Will Your Initial and Full Market Rollout Look Like?

Ask yourself the following questions to develop a robust go-to-market strategy:

a. What is your initial rollout plan? What is the expansion plan or full market rollout plan? Define the initial rollout including trials to engage friendly users or a small set of customers. Offer incentives if they need it.

b. What is your full business case? Leveraging the pricing, cost, and revenue projections you have decided on earlier, develop a version of your business case to help you understand the projected financial health of the organization per a given region and given period.

c. What are the price points that you will initially rollout with and when will you consider changing them? How are you going to measure customer response to your price points?

d. What are your target markets? Where will you offer your product to customers and in what markets and segments?

e. How fast will you expand to include additional markets and customer segments? During the full market rollout, is a trial needed in every new market, or would the results from previous trials be sufficient to prove the value to customers in each new region?

f. What is your customer acquisition process? How are you going to sell your product when ready? Define how customers are going to acquire and receive it i.e., what is the customer buying process? Would the business need a physical presence, online presence, partnership channels, or a hybrid model?

g. Do you need a channel partner? If so, what are the criteria that will help you in selecting one and in what markets? Also, what terms and conditions, including revenue sharing, are you willing to offer to the partner, and who will service and support the acquired customers?

h. What is your marketing strategy? Where are the best places to market the business? How will people find out about your product? What marketing channels are you going to use?

i. What is your sales strategy? Who will sell your products and what is the structure of this team? What is your sales process and how long is your expected sales cycle? Do you have legally reviewed and approved customer agreements with terms and conditions defined and ready to use when closing a deal?

j. What is your post-sales support and maintenance plan, and what are your planned customer service operations? How will you address urgent and critical product issues vs. non urgent raised issues? For business-to-business sales, your commitment will need to be described in a service level agreement (SLA), which should be reviewed by a professional legal expert, and which should reflect your terms for response time to the delivery, installation, and for solving product issues.

k. For digital products, what is your cloud-hosting strategy? Will you be hosting your application at a highly-reliable commercial cloud company, within your own servers, or installed within each customer's servers?

l. How do you measure success? Define your key success indicators, including customer satisfaction, sales, customer acquisition, customer engagement, repeat sales, and growth milestones, to keep aware of what works best and what needs to change. Understanding the performance of your GTM helps you optimize and adjust it as you go along; for example, your spending on various media channels.

10. Secure Your Funding: What Capital Do You Need and How Will You Find Investors?

Ask yourself the following questions to ensure you are prepared when reaching out for external funding:

a. How are you going to fund the business's initial launch and its operational expenses until it can stand on its own?
b. How much money does the business need to become customer ready or reach its first milestone?
c. How much incremental funding would be needed over the course of the next six months to a year, or until reaching the breakeven point milestone?
d. When do you need to raise funding? Understand your timeline in order to start early enough in the process. The funding process can take significant effort, time, and several cycles, and it's a strain on the leader and their team. You must have the necessary capital to operate during the time you are seeking external funding.
e. How much capital have the leaders and/or other investors paid into the company already?

f. Do you have a valuation for the company? If you decided to invite investors to join, could the value be shared with them? Can you justify the determined valuation with hard facts, such as sales, revenue, growth trends, customers registration, or other market data?

g. When funding is needed, which source should you consider?

i. **Self-Funding:** Can the business founder self-fund the business? If so, how much are you willing to spend to kick off the company? You should consider your own living and family obligations and risk level while determining the amount you can allocate as an investment.

ii. **Customer Funding:** Can your business generate sufficient sales to cover your initial cost? The best funding source for a business is their own customers, i.e., from its own sales and operations. In some scenarios, customers may prepay for the product or the service. Some may even be willing to contribute an amount as an investment if they see a strategic advantage to their business.

iii. **Bank Lending:** Do you need the funds to support the delivery of a guaranteed contract? If so, a bank loan may be a good avenue to help you execute the contract instead of selling equity. Banks are more willing to offer a loan when there is a guaranteed income.

iv. **Friends and Family:** Could friends and family invest? Consider them first during the early stage of your venture. However, do not borrow an amount that can alter the lives of those who are close to you if the risk is high.

v. **Angel Investors:** Could you approach angel investors? They can provide the initial needed funds for early-stage activities and the process can help you, as the founder, to learn more about the perceived risk factors through the eyes of unbiased and experienced investors.

vi. **Crowdfunding:** Are there crowdfunding platforms that would suit your product or business? For some, these may offer the kick start they need. Not all platforms are equal, and some serve different purposes. Be sure you are comfortable with the fees and terms associated with platform usage.

vii. **Incubators and Accelerators:** Can you consider incubators or accelerators? They can offer valuable connections to potential investors, partners, and customers for friendly trials. They can also be a helpful bridge to larger funding institutions. But be sure to carefully select who to engage by researching their performance and talking with them directly and with others that have engaged with them.

viii. **Venture Capital:** Are there venture capitalists and corporate institutions which may want to invest in your business? If your company has demonstrated market success and customer traction and still requires funds to meet additional milestones, then a VC can be a good option. Remember that they invest not only in products, but also in the team.

11. Prepare Your Pitch: How Can You Convince Investors, Partners, and Customers to Get on Board?

Ask yourself if you have these pitches below ready. Have you already tested them with friendly audiences, such as team members in your company, advisors, or mentors, before delivering them to external customers, investors, or the public in general?

    a. What is your five-second pitch? (one – two sentences outlining your value proposition)
    b. What is your elevator pitch? (30 – 90 seconds)
    c. What is your detailed pitch? (five – 15 minutes)
        i. To customers
        ii. To investors

Example of a Detailed Investor Pitch

Below is an example of a good flow for an investor pitch with the purpose of generating interest to invest funds in exchange for a stake in the company:

1. **The team.** Present the team and highlight their key strengths. Establish trust and confidence in their capabilities, skills, and experience related to the business. The audience will also be gaining insight into how well the team can communicate and how easy it is to work with them in the future.

2. **The problem.** Describe the pain point or the problem that the business has identified; a need to be solved and addressed. This should include why it's important to solve and how urgent it is.

3. **The customer.** Discuss who the target customer segments are and why they will care about the product. How are they addressing the problem today without the product and why are existing solutions not fulfilling this need? Support the discussion with facts and data.

4. **The mission.** Describe the mission statement and what your company wants to do about the problem.

5. **The solution.** Present your solution, product, or service that addresses this opportunity. This should include a brief description of its components and how it works.

6. **Differentiation.** What makes it unique and different from other products in the market? Why would someone choose your solution over its competitors?

7. **Competition.** Who are the competitors and how do you stack up against them? What are their strengths and weaknesses in comparison to yours? You may choose to perform a SWOT analysis or a similar tabular matrix.

8. **Finances.** Demonstrate your knowledge by presenting your key numbers. Describe your financials, including pricing, cost, the opportunity size over the next one to three years, number of customers, and associated revenue you are capturing or want to capture. You may want to add a longer projection period such as five years. The key here is to be able to describe the projections based on reasonable and believable assumptions. Use a bottom-up approach that demonstrates customer growth month over month, units sold with prices, and expenses on a monthly or quarterly basis. This is more accurate than a top-down model where you assume a percentage gain from the overall market size; this simply demonstrates the opportunity potential.

9. **Go-to-market strategy.** Describe your sales and go-to-market strategy. Discuss how you are going to achieve the projected finances in order to make those projections achievable and credible.

10. **Funding.** Describe the level of funding you need, what has been received so far, if any, from previous funding events or rounds, including any self-funding. Discuss what you need moving forward, the instrument of finance that you are proposing, such as straight capital in exchange for equity, convertible notes, a loan, etc. Outline what the funds are going to be used for, and what milestones you will hit by what timeframe. For example, funds might be used for a market trial, to grow the MVP, to deploy the product in a particular city or country, to expand a market, to hire skilled employees in a specific capacity, to grow revenue, or to build and expand your competitive advantage.

Example of a Detailed Sales Pitch with a Potential Business Customer

Always be prepared. Learn about the customer, their business, and their problems ahead of time. Attempt to identify their needs and how your solution or offering may solve those needs. When you meet, listen to the customer. You may start the meeting by asking them, before your pitch, if they're willing to discuss their current priorities and the pain points they're facing. The research you did before meeting can help significantly, including identifying the key team members that should be in the meeting, including decision makers.

The presentation should start by establishing credibility in your business and product and then moving on to discuss the solution and how it addresses the customer's needs. An example flow is as follows:

1. **Company.** Describe the company's history, strengths, and existing customers. Establish credibility and trust in the company by expressing your understanding of the market, the industry, and outlining your success stories so far.

2. **State the problem.** Describe the opportunity or the pain points as you see them in the market and confirm if the customer is experiencing the same problems. Quantify the problem if possible and how impactful it is on them or their customers, perhaps by discussing revenue or any other metrics that are important to them. While the solution you intend to present may address or solve many problem types, be sure you state the problem that is specifically related to the customer, and not in general.

3. **The mission.** Describe the mission statement and how this captures what your company is doing about the problem.

4. **The solution.** Present your solution, product, or service that addresses this opportunity. Give a brief description of its components, its architecture if applicable, and how it works. Discuss how it is implemented and used by the customer to resolve the pain point, or how it offers value and benefits.

5. **Implementation.** Discuss what it would take for this product to be acquired by the customer. Highlight the process to access the offering. It may be as simple as setting up online access, or it may require physical installation and integration.

6. **Demonstration (demo).** If possible, provide a real-time demonstration of the product, how it's used, and highlight the ease of use, features, and benefits. This is a key step where you

have the chance to showcase your offering and impress the customer.

7. **Differentiation.** Discuss what makes it unique and different. Explain not only your competitive advantage, but also how the customer will gain a unique edge if they use your offering.

8. **Advantages and market results.** Discuss the advantages it provides and present real data if available. You may use data from friendly customers who used your offering and provided you with permission to share. Data from previous trials with users may also be used.

9. **Pricing.** Depending on the maturity of the relationship, the stage of the sales process, and the type of offering, you may choose to present the pricing structure, or pricing options, especially if you feel that your pricing strategy is an added competitive advantage.

10. **Feedback and follow ups.** Ask for customer feedback and propose a call to action follow up. Some customers may offer feedback throughout the presentation, while others may choose to wait until the end. Allow time for questions, and clarifications. Ideally, you may arrange a call to action, such as a follow up with the same team, or other key members that are not present. Try to learn who the decision makers and influencers are and what their typical purchasing process is. Based on their feedback and input, you can structure these next steps to advance the sales process.

12. Evaluate Your Risk: What is the Likelihood of Failure and How Will You Prepare for and Mitigate Your Business Risk?

Ask yourself the following questions to help prepare for the worst and thus increase your chances of success. What can make my business fail? Identify the top three to five factors and have a plan to mitigate them.

   a. What if customers don't come and don't need my offering? Don't miss the mark on this one. Validate their interest and the value of your offering before you go deep into your new venture. Go back to Principle 1.
   b. What if customers find my product too expensive? Examine your messaging, your value proposition, and differentiation. Research your competition's pricing and understand your cost structure to determine your pricing flexibility in offering new models that are more acceptable to customers.
   c. What if new competitors emerge within a short period of launch, or existing competitors evolve to present a similar value proposition?
   Be prepared for the financial impact and add a risk margin to your revenue projections. Be ready to compete in differentiating your business, and be prepared to grow your market advantage through your product's roadmap, marketing, and in your passion to satisfy your customers.
   d. Will the business have sufficient funds to launch and operate until it can be fully self-sufficient?
   Understand your capital needs and don't assume the best-

case scenario. Have a plan to beat your market goals but be prepared for falling short. Identify capital sources that you can utilize, and understand the time required to access it should you need it.

e. Are there measures in place to track progress?
Understand your performance against important financial and operational measures to evaluate the health of your business, the level of risk, and how close you are to meeting your goals. When you recognize a deviation in its early stages, quick action can be taken to correct.

f. What can be learned from other companies which launched similar products previous to yours?

 i. Have others with a business concept similar to yours struggled or failed before you? Why do you think they failed?

 ii. Have others tried it before you and succeeded? Why do you think they succeeded?